"Brassinosteroids: Practical Applications in Agriculture and Human Health"

Edited By

Adaucto Bellarmino Pereira-Netto

Department of Botany-SCB, Paraná Federal University, Curitiba-PR

BRAZIL

CONTENTS

FOREWORD

Brassinosteroids (BRs) are growth-promoting hormones that occur endogenously across the plant kingdom. Among plant hormones, BRs are structurally the most similar to animal steroid hormones, which have well-known functions in regulating embryonic and post-embryonic development and adult homeostasis, and like their animal counterparts, BRs regulate multiple physiological processes essential to both somatic and reproductive development. After their discovery in the 1970s, an array of experiments including bioassays, greenhouse and field trials, and physiological tests involving exogenous application of BRs, suggested these newly discovered plant compounds influenced organ elongation, leaf morphogenesis, senescence, male fertility, pollen development and fruit ripening. The biochemical delineation of the BR biosynthetic pathway and the discovery of BR deficient and insensitive mutants in *Arabidopsis thaliana* and several crop plants in the 1990s provided convincing evidence that BRs were as essential for normal plant development as the better-known plant hormones such as auxins, cytokinins and gibberellins.

Numerous books and reviews covering currently known aspects of BR biology and chemistry are available, including biosynthesis and metabolism, physiological effects and signal transduction. A great deal of progress has been made recently in understanding specific components of BR signaling and in clarifying mechanisms by which BR perception ultimately results in regulation of large gene networks associated with numerous developmental programs. The number of physiological processes known to involve BR action has also expanded. In parallel with these advances in molecular biology and physiology, a large group of researchers have continued to demonstrate the utility of applying BRs to a broad spectrum of problems in agriculture, environmental science and even medicine. It is these practical applications of BRs that is the content of the current volume and this book occupies a unique niche in BR literature by focusing exclusively on practical topics.

Given the large body of literature on application of BRs to agricultural crops and the demonstration that even minimal modulation of BR signaling or biosynthesis can have dramatic effects on rice yields and tomato fruit development, it is now evident that BRs will have practical applications in regulating the growth and development of many crop species. It is well known that BR field applications yield the most visible responses when crops are under stress, and three chapters in the current volume cover abiotic stresses such as drought, salinity and heavy metals; along with biotic stresses such as pathogen attack and plant defense responses. The role of plant hormones, including BRs, in regulating plant shoot growth is also well known and an additional chapter discusses novel uses of BRs in micropropagation. The final three chapters introduce the fascinating literature that shows the potential for BRs and their analogs in inhibiting human viral pathogen replication and the possible role of BRs and analogs as anti-inflammatory, anticancer and antiproliferative agents.

The editor and authors of this volume should be commended for adding a unique contribution to BR literature that will be appreciated by those directly working on practical applications of BRs as well as molecular biologists studying BR mechanisms who wish to become familiar with the range of applications being considered for this family of plant steroids. Furthermore, medical researchers studying the role of phytochemicals in addressing treatment of human diseases will find the final three chapters of significant interest.

Steven D. Clouse
North Carolina State University
Raleigh, NC USA

PREFACE

Brassinosteroids (BRs) comprise a specific class of low-abundance, natural polyhydroxy steroidal lactones and ketones. These steroids of ubiquitous occurrence in plants are known for a long time to induce a broad variety of responses in plants such as cell expansion and division, shoot elongation, bending, resistance against biotic and abiotic stress, and reproductive and vascular development. However, more recently, these compounds have also been shown to present antiviral, anti-inflammatory and anticancer activity in animal systems.

This e-Book provides and up-to-date view of practical applications for BRs. Each chapter presents a review of the available literature on various aspects of the use of BRs for the improvement of human welfare. The first chapter provides an overview about the increase in plant productivity and improvement of human health, besides addressing bio-safety issues. Second chapter describes the mitigation of water stress and saline stress. Third chapter uncovers the improvement of protocols for micropropagation. Fourth chapter will discuss anti-herbivory properties. Chapter five provide an insight into the enhancement of tolerance to heavy metals. The following chapters will deal with the use of BRs for the improvement of human health. Chapter six will describe antiviral properties. Chapter seven summarizes antiherpetic and anti-inflammatory activities. Finally, Chapter eight will describe anticancer activities. The editor hopes that this e-Book will provide valuable and updated information on practical applications for BRs in agriculture and human health, besides leading to new research to expand the use of BRs for human welfare.

The editor wish to thank all of the contributors for their willingness to provide the highest quality, state-of-the-art contributions to this e-Book.

Adaucto Bellarmino Pereira-Netto
Paraná Federal University
Brazil

List of Contributors

A. Ahmad

Department of Applied Sciences, Higher College of Technology, Al-Khuwair, Sultanate of Oman.

L. E. Alché

Laboratorio de Virología: Agentes antivirales y citoprotectores. Departamento de Química Biológica. Facultad de Ciencias Exactas y Naturales, Universidad de Buenos Aires, Ciudad Universitaria, Pabellón 2, Piso 4, C1428EGA, Buenos Aires, Argentina.

R. Bhardwaj

Department of Botanical and Environmental Sciences, Guru Nanak Dev University, Amritsar 143005, Punjab, India.

M. L. Campos

Department of Biological Sciences (LCB), Escola Superior de Agricultura "Luiz de Queiroz" (ESALQ), Universidade de São Paulo (USP), Piracicaba-SP, Brazil.

V. Castilla

Laboratorio de Virología. Departamento de Química Biológica. Facultad de Ciencias Exactas y Naturales, Universidad de Buenos Aires, Ciudad Universitaria, Pabellón 2, Piso 4, C1428EGA, Buenos Aires, Argentina.

Q. Fariduddin

Plant Physiology Section, Department of Botany, Aligarh Muslim University, Aligarh-202 002, India.

L. R. Galagovsky

Department of Organic Chemistry and UMYMFOR (CONICET-FCEyN), School of Exact and Natural Sciences, University of Buenos Aires, Argentina.

N. Handa

Department of Botanical and Environmental Sciences, Guru Nanak Dev University, Amritsar 143005, Punjab, India.

S. Hayat

Plant Physiology Section, Department of Botany, Aligarh Muslim University, Aligarh-202 002, India.

L. Hoffmannová

Laboratory of Growth Regulators, Faculty of Science, Palacký University & Institute of Experimental Botany ASCR, Šlechtitelů 11, Olomouc, 783 71, Czech Republic.

M. Kanwar

Department of Botanical and Environmental Sciences, Guru Nanak Dev University, Amritsar 143005, Punjab, India.

D. Kapoor

Department of Botanical and Environmental Sciences, Guru Nanak Dev University, Amritsar 143005, Punjab, India.

L. Kohout

Laboratory of Growth Regulators, Faculty of Science, Palacký University & Institute of Experimental Botany ASCR, Šlechtitelů 11, Olomouc, 783 71, Czech Republic.

Z. Kolář

Laboratory of Molecular Pathology, Department of Pathology, Faculty of Medicine, Palacký University, Hněvotínská 3, 775 15 Olomouc, Czech Republic.

F. M. Michelini

Laboratorio de Virología: Agentes antivirales y citoprotectores. Departamento de Química Biológica. Facultad de Ciencias Exactas y Naturales, Universidad de Buenos Aires, Ciudad Universitaria, Pabellón 2, Piso 4, C1428EGA, Buenos Aires, Argentina.

J. Oklešťková

Laboratory of Growth Regulators, Faculty of Science, Palacký University & Institute of Experimental Botany ASCR, Šlechtitelů 11, Olomouc, 783 71, Czech Republic.

A. B. Pereira-Netto

Department of Botany-SCB, Paraná Federal University, Curitiba-PR, Brazil.

L. E. P. Peres

Department of Biological Sciences (LCB), Escola Superior de Agricultura "Luiz de Queiroz" (ESALQ), Universidade de São Paulo (USP), Piracicaba-SP, Brazil.

J. A. Ramirez

Department of Organic Chemistry and UMYMFOR (CONICET-FCEyN), School of Exact and Natural Sciences, University of Buenos Aires, Argentina.

I. Sharma

Department of Botanical and Environmental Sciences, Guru Nanak Dev University, Amritsar 143005, Punjab, India.

J. Steigerová

Laboratory of Molecular Pathology, Department of Pathology, Faculty of Medicine, Palacký University, Hněvotínská 3, 775 15 Olomouc, Czech Republic.

M. Strnad

Laboratory of Growth Regulators, Faculty of Science, Palacký University & Institute of Experimental Botany ASCR, Šlechtitelů 11, Olomouc, 783 71, Czech Republic.

B. V. Vardhini

Department of Botany, Telangana University, Dichpally, Nizamabad -503322, Andhra Pradesh, India.

M. B. Wachsman

Laboratorio de Virología. Departamento de Química Biológica. Facultad de Ciencias Exactas y Naturales, Universidad de Buenos Aires, Ciudad Universitaria, Pabellón 2, Piso 4, C1428EGA, Buenos Aires, Argentina.

M. Yusuf

Plant Physiology Section, Department of Botany, Aligarh Muslim University, Aligarh-202 002, India.

CHAPTER 1

Current Scenario of Applications of Brassinosteroids in Human Welfare

Renu Bhardwaj[*], Indu Sharma, Mukesh Kanwar, Neha Handa and Dhriti Kapoor

Department of Botanical and Environmental Sciences, Guru Nanak Dev University, Amritsar 143005, Punjab, India

Abstract: Brassinosteroids (BRs) regulate plant growth and development and show structural similarities to animal steroidal lactones. They are ubiquitously distributed throughout plant kingdom and regulate a broad spectrum of plant developmental and physiological processes, including gene expression, cell division and expansion and differentiation at nanomolar to micromolar concentrations. Exogenous applications of BRs revealed their enhancing effects on yield and quality of crops, vegetables and fruits. Besides, BRs have also been reported to play a significant role in stress-protection in both biotic and abiotic stress in plants. Recently, BRs have attained worldwide attention for their bioactivities in diverse test systems as well as in agricultural applications. Burgeoning studies have divulged antiviral, antifungal, antiproliferative, antibacterial, neuroprotective and immunomodulatory properties of BRs in animal systems. In human cells, BRs and their analogues are reported to inhibit cell growth in cancer cell lines. Keeping in view the agricultural, stress protective and medicinal properties of BRs, they are emerging as potential candidates for use in human welfare.

Keywords: Antiviral, agriculture, brassinolide, growth, immunomodulatory, phytohormones, stress protection.

INTRODUCTION

Steroids, an essential group of hormones in metazoans, act as imperative signaling molecules which control a wide array of growth and physiological processes. Nonetheless, steroidal sex hormones like estrone, testosterone, progesterone, corticosteroids and several analogues of these compounds had also been reported from plants [1]. Exogenous application of steroidal sex hormones on plants has been ascertained to be exceedingly effectual in terms of significant morphological and physiological changes [2-7]. In late 1970's, brassinosteroids (BRs) has been discovered as a group of novel plant steroid hormones having structural similarity to animal steroid hormones. Mitchell *et al.* [8] reported promotion in stem elongation and cell division by the treatment of organic extracts of *Brassica napus* pollen. The first BR was isolated in a crystalline form from rape pollen by Grove *et al.* [9] and was named brassinolide (BL). The general structure of BRs is characterized by a carbon skeleton with four fused rings. Different BRs vary by the functional groups attached to these rings and length of side chain. Yokota *et al.* [10] isolated the second BR which is known as castasterone (CS). Since then, about 70 BRs (65 unconjugated and 5 conjugated) have been reported in plants [11-14]. Besides the known BRs, there might be additional undiscovered BRs and their conjugates in plants [15].

BRs are distributed ubiquitously throughout the plant kingdom and had been isolated from 58 plant species (49 angiosperms, 6 gymnosperms, a pteridophyte - *Equisetum arvense*, a bryophyte - *Marchantia polymorphha* and an alga - *Hydrodictyon reticulatum*). Endogenous levels of BRs vary from tissue to tissue and with the age of the plant. However, at nanomolar and micromolar concentrations they have been reported to be very effective [13]. Pollens and immature seeds are found to have the highest concentrations of BRs [16]. Like their animal counterparts, BRs also control a wide spectrum of developmental and physiological processes in plants, including regulation of gene expression, cell division and expansion, germination, vegetative and reproductive development, vascular differentiation, root growth, programmed cell death, and homeostasis [5, 6, 13, 17-19]. At cellular level, BRs stimulate elongation and fission,

*****Address correspondence to Renu Bhardwaj:** Department of Botanical and Environmental Sciences, Guru Nanak Dev University, Amritsar 143005, Punjab, India. E-mail: dr.renubhardwaj@gmail.com

Adaucto Bellarmino Pereira-Netto (Ed)

activate protein and nucleic acid synthesis, enhance photosynthetic capacity, alter mechanical properties of cell wall and permeability of cell membranes [16, 20, 21]. At the whole plant level, BRs promote overall growth, reproductive development, shorten the period of vegetative growth, increase crop yield and improve the quality of fruits [16, 22-24]. In addition to their growth regulatory activities, BRs have been explored for their dynamic role against biotic and abiotic stress-protection in plants [25-29]. Exogenous application of BRs stimulated inner potentials of plants that are helpful not only in better survival in stress conditions, but also in receding biotic stress caused by pathogens such as viruses, fungi and bacteria [30-34]. Due to active involvement of BRs in a large number of physiological processes, these are anticipated as hormones of the 21st century [16].

BRs are synthesized from phytosterols through complex oxidative reactions resulting in the synthesis of the biologically most active BR, *i.e.* brassinolide (BL) [35]. *In vivo* studies revealed two alternative biosynthetic pathways, the early and late C-6 oxidation pathways, which utilize 6-oxo and 6-deoxo intermediates, respectively [35-37]. The phenotypes of the BR-deficient mutants indicated that BRs are essential hormones and take part in light-regulated development. Attempt to identify BR-insensitive mutants of *Arabidopsis* resulted in the isolation of the dwarf mutant named *bri1* [38, 39] which was found to have strong homology to leucin rich receptor kinases. Now, numerous BR biosynthetic enzymes have been identified through the characterization of BR-deficient mutants in *Arabidopsis* [35, 40]. Moreover, *Arabidopsis* mutants helped in the identification of signaling proteins involved in BR signal transduction pathway [40]. Genetic screens and proteomic approaches revealed that the BR signal transduction cascade includes BR perception by the BR INSENSITIVE1 (BRI1) receptor kinase, at the cell surface, and activation of BRI1/BRI1-ASSOCIATED RECEPTOR KINASE 1 (BAK1) kinase complex by transphosphorylation. Subsequently, BRASSINOSTEROID INSENSITIVE 2 (BIN2) kinase is dephosphorylated and inactivated resulting in the accumulation of unphosphorylated BRASSINAZOLE RESISTANT (BZR) transcription factors in the nucleus [40].

ROLE OF BRASSINOSTEROIDS IN STRESS MANAGEMENT

Plants are concurrently exposed to a range of abiotic and biotic stresses rather than a particular stress condition, which are most lethal to crops [41]. Exogenous BR application can ameliorate both biotic and abiotic stresses in plants, thereby increasing their tolerance to adverse conditions.

Brassinosteroids and Biotic Stress Management

Exogenous applications of BRs resulted in enhanced resistance to various biotic stresses caused by pathogens such as viruses, fungi and bacteria [19, 30-34]. BRs are resorted to enhance plant resistance against fungal pathogen infections in potato plants [16, 42]. BRs treatments significantly reduced *Magnaporthe grisea* and *Xanthomonas oryzae* induced rice blast and bacterial blight respectively [30]. The Herpes Simplex Virus Type 1(HSV-1) and Arena Virus replication in cell culture was inhibited by treatment of BRs and their synthetic derivatives [34, 43]. In addition to antiherpetic activity effects of BRs on insect development, particularly on molting, were reviewed by Zullo and Adam [44]. Treatment of BRs to root knot nematodes (*M. incognita*) also stimulated their antioxidative defence system [45] as well as egg hatching and juvenile emergence in [31].

BRs are the first phytohormones which are true antiecdysteroids and they show striking structural similarities to the ecdysteroids. Due to structural resemblance of BRs with ecdystriods, they have been widely used to delay insect molting. They are supposed to compete with natural hormone for hormone receptor binding site [46]. Two triterpenoids isolated from seeds of cruciferous plants, cucurbitacins B and D, were found to be insect steroid hormone antagonists acting at the ecdysteroid receptor [47]. These reports suggest the possible implications of BRs in insect pest control. Another similarity in steroid responses between plants and insects is programmed cell death. Ecdysteroids trigger the massive death of larval tissues during the early stages of metamorphosis, ridding the animal of these obsolete tissues to make way for their adult counterparts [48]. This response has been extensively studied in *Drosophila* and shown to occur by autophagy with hallmark features of apoptosis, including DNA fragmentation and caspase activation [49-51]. There is evidence that BRs induce programmed cell death during xylogenesis. The specialized xylem vessels that conduct water through plants are made up of individual dead cells called tracheary elements [52]. The role of BRs in apoptosis and programmed cell death make them potential candidates in the management of lifecycles and different developmental stages of insects and nematodes which spoils the agricultural crops, thereby, helping in sustainable agriculture.

Brassinosteroids and Abiotic Stress Management

In the third world countries, major crop loss is due to the environmental or abiotic stresses. Stress protective action of BRs is the result of a complex sequence of biochemical shifts, such as activation or suppression of key antioxidant enzymatic reactions, induction of protein synthesis, and the production of various chemical defence compounds. BRs have been widely reported to confer abiotic stress resistance in plants suggesting their essential necessity for normal functioning of the plant defense system. They are reported to have stress protective effects in a wide range of plants such as *Arabidopsis thaliana, Brassica napus, B. juncea, Cicer arietinum, Raphanus sativus, Zea mays, Oryza sativa etc.* against a plethora of environmental stresses *viz.* drought, extreme temperatures, heavy metals, herbicidal injuries and salinity [27, 53-57]. In our earlier studies, we have reported that 24-Epibrassinolide (EBL) and 28-Homobrassinolide (HBL) treatments (presowing) improved the shoot emergence and plant biomass production in *Brassica juncea* seedlings and plants under heavy metal stress (Cu, Zn, Mn, Co and Ni). EBL and HBL have also been found to reduce the heavy metal uptake and accumulation in *B. juncea* seedlings and plants [12, 58-61]. In rice, applications of BRs reduced the damage caused by pesticides (simazine, butachlor, or pretilachor) [62]. The protective effects of brassinosteroids and its structural analogs had been studied on growth, lipid peroxidation and antioxidative system of rice, soyabean, tomato, potato and *Cicer arietinum* under various stresses and it was observed that application of these hormones lowered the oxidative stress and promoted plant growth [63-65]. Further, BRs have been reported to increase the activities of antioxidant enzymes of the plants under heavy metals stress [56, 66-76].

ROLE OF BRASSINOSTEROIDS IN MEDICINAL APPLICATIONS

Recent studies on the biological activities of BRs in various animal test systems depicted their antibacterial, anticancerous/antiproliferative, antifungal, antigenotoxic, antiviral and ecdysteroidal properties. Consequently, BRs have the prospect as a potential future medicine for treating cancer, fungal, bacterial and viral infections, among others [30-34, 77, 78]. Some of the recent studies done in this direction are as follows:

Antibacterial/Antiviral Activities of Brassinosteroids

BRs and their synthetic derivatives/analogues are also reported as inhibitors of Herpes Simplex Virus Type 1 (HSV-1) and Arena Virus replication in cell culture [43, 79 and Chapter **6** in this book]. Synthetic methods to obtain BRs analogues and their *in vitro* antiviral activity against RNA and DNA viruses were described by Wachsman *et al.* [34]. Some of these analogues confirmed good antiviral activity against Junin Virus (JV) (*Arenaviridae*), Measles Virus (MV) (*Paramixoviridae*), Herpes Simplex Type 1 and 2 (HSV-1 and HSV-2) (*Herpesviridae*) [34]. However, viruses respond to antiviral treatment with a rapid selection of drug resistant mutant particles, compelling virologists to search for new active compounds. Animal viruses tested for their susceptibility to BRs analogues comprised two RNA monocistronic viral families, *Paramyxoviridae* and *Arenaviridae,* and one DNA virus family *Herpesviridae*, all of them important human pathogens.

In vitro and *in vivo* antiherpetic activity of three new synthetic BRs analogues *viz.* (22S,23S)-3β-bromo-5α,22,23-trihydroxystigmastan-6-one, (22S, 23S)- 5α-fluoro-3β-22,23-trihydroxystigmastan-6-one, (22S, 23S)-3β-5α,22,23-trihydroxystigmastan-6-one in human conjunctive cell lines (IOBA-NHC) as well as in the murine Herpetic Stromal Keratitis (HSK) experimental model were reported by Michelini *et al.* [80]. These compounds prevented the multiplication of HSV-1 in NHC cells in a dose dependent manner, when added after infection, with no cytotoxicity. *In vitro* studies had confirmed the ability of EBL to arrest or reduce the growth of the HIV in cultured infected cells, further strengthening potential of BRs in the cure or prevention of HIV infection and related conditions (AIDS related complex) [81]. The antiviral effects of a synthetic BR ((22S, 23S)-3β-bromo-5α,22,23-trihydroxystigmastan-6-one) against replication of Vesicular Stomatitis Virus (VSV that causes an economically important disease in cattle, horses and swine) in vero cells has been reported [32]. An ocular chronic immunoinflammatory syndrome named herpetic stromal keratitis that might lead to vision impairment and blindness in mice is caused by HSV-1 and Michelini *et al.* [82] has reported antiherpetic activity of synthetic BRs against HSV-1 (Chapter **7**).

Anticancerous/Antiproliferative Activities of Brassinosteroids

Recently, BRs have been reported for their anticancerous and antiproliferative prospective in various test systems which can be used for the development of new BR-derived generation of anticancer drugs [33, 77, 83]. EBL treatments (10^{-16} to 10^{-9} mol.l^{-1}) modified the growth and production characteristics of a mouse hybridoma. EBL resulted in increase in the value of mitochondrial membrane potential, drop of intracellular antibody level, increase in the fraction of cells in the G_0/G_1 phase, and decrease in the fraction of cells in the S phase. BRs also been shown to affect the viability, proliferation, differentiation, apoptosis and expression of some cell cycle related proteins in cancer cell lines [33]. Cytotoxic activity of BRs was tested *in vitro* by Calcein AM assay whereas TUNEL, DNA ladder assay and immunoblotting techniques were used for the analysis of changes in cell viability, proliferation, differentiation and apoptosis. 28-homocastasterone (28-HCS) inhibited the viability of human breast adenocarcinoma cell lines (MCF-7–estrogen-sensitive, MDA-MB-468–estrogen-insensitive), human acute lymphoblastic leukemia cell line (CEM) and human myeloma cell line (RPMI 8226). Also, significant reduction or induction in the expression of *p21*, *p27*, *p53*, cyclins, proteins of the Bcl-2 family and ER-alpha was observed. Malíková *et al.* [77] tested the effects of 28-HCS and EBL on the viability, proliferation, and cycling of hormone-sensitive/insensitive breast (MCF-7/MDA-MB-468) and prostate cancer cell lines (LNCaP/DU-145) to determine whether the discovered cytotoxic activity of BRs could be, at least partially, be related to BR-nuclear receptor interactions. Both BRs inhibited cell growth in a dose dependent manner in the cancer cell lines. Flow cytometry analysis showed that BR treatment arrested, MDA-MB-468, LNCaP and MCF-7 cells in G1 phase of the cell cycle and induced apoptosis in MDA-MB-468, LNCaP, and slightly in the DU-145 cells. These results proved that natural BRs can inhibit growth, at micromolar concentrations, of several human cancer cell lines without affecting the growth of normal cells.

Wachsman *et al.* [79] reported that some natural brassinosteroids (28-homocastasterone, 28-homobrassinolide) and their synthetic analogues have *in vitro* antiviral activity against several pathogenic viruses, such as Herpes Simplex Virus Type 1 (HSV-1), Arena Viruses and Measles Virus (MV). Several analogues of BRs have been shown to be 10 to 18-fold more active than ribavirin (used as the reference drug) for HSV-1 and Arena Viruses. However, further studies are needed to define the precise *in vitro* antiviral mechanism of these BR analogues and to correlate molecular structure and bioactivity. There is also one report describing possible effects of EBL on cultured mouse hybridoma cells [84]. Typical effects of these compounds are: (a) increase in the value of mitochondrial membrane potential, (b) decrease of intracellular antibody level, (c) increase in the fraction of the cells in G0/G1 phase, and (d) vice versa decrease of S-phase cells. Furthermore, the density of viable cells was significantly higher at EBL concentrations of 10^{-13} mol/L and 10^{-12} mol/L [83].

Oklestková *et al.* [84] have found that a series of natural BRs are effective in growth inhibition of many different cancer cell lines at micromolar concentrations despite of their minimum effects on normal cells. Their cytotoxic activity could be, at least partially, related to interactions with steroid receptors. Application of BRs induced cytotoxicity and growth inhibition of breast and prostate carcinoma cells. Natural BRs are useful especially for treating disorders, some of them involving cell proliferation, and including cancer, Alzheimer disease, Huntington disease, steroid-induced osteoporesis, sexual differentiation disorders, hyperadrenocorticism associated with sex steroid excess, androgen insensitivity syndrome, glucocorticoid insensitive asthma, steroid-induced cataract, and deficiency of P450 oxidoreductase. Oklestková *et al.* [84] discussed techniques which could arrest of the cell cycle by natural BRs resulting in apoptotic changes in cancer cells, thus recommended the use of BRs for treatment of the adverse effects of hyperproliferation on mammalian cells *in vitro* and *in vivo*, especially treatment of hyperproliferative diseases in mammals. Also, described new use for treating consisting in a new therapeutic way for modifying cell viability of human breast and prostate cancer cells. Hence, they can be used as antimitotic and apoptotic drugs, particularly as anticancerous drugs.

ROLE OF BRASSINOSTEROIDS AGAINST HERBICIDES AND PESTICIDES

Herbicides and pesticides are major pollutants of worldwide concern. These are persistent compounds which get accumulated in the environment and are non-biodegradable pollutants therefore, contaminate the ecosystem and enter food chains. Thus, they are bioaccumulated and biomagnified at higher trophic levels and also affect human health directly or indirectly [85]. Certain pesticides have mutational effects on human DNA molecules. DDT which is commonly used is reported to have lethal effects in small doses. It is also suspected to be carcinogenic in human tissues. Various pesticides such as DDT, DDE, DDD, dieldrin, heptachlor epoxide and numerous herbicides such as 2,3,5-T (2,3,5-trichlorophenoxy acetic acid) dioxin have been extensively used for control of diseases and crop destroying insects. BRs diminish herbicidal injury to rice caused by simazine, symetrin, butachlor and pretilachlor [86], perhaps by reducing transpiration and herbicidal absorption and by counteracting the herbicide-induced inhibition of photosynthesis [14]. However, the mechanism of BR action has not been studied.

Applications of BRs are known to protect crops from the toxicity of herbicides, fungicides and insecticides [87], thus may prevent the entry of BRs into food chain. Therefore, BRs can become promising, environment friendly, natural substances suitable for wide application to reduce the risks of human and environment exposure to pesticides. Application of EBL accelerated metabolism of various pesticides and consequently reduced their residual levels in cucumber (*Cucumis sativus* L) [87]. Chlorpyrifos, a widely used insecticide, caused significant reductions of net photosynthetic rate (Pn) and quantum yield of PSII (Φ_{PSII}) in cucumber leaves. EBL pretreatment alleviated the declines of Pn and Φ_{PSII} caused by chlorpyrifos application, and this effect of EBL was associated with reductions of chlorpyrifos residues. EBL enhanced the expression of *P450* and *MRP*, which encode P450 monooxygenase and ABC-type transporter, respectively. The stimulatory effect of EBL on pesticide metabolism was also observed for cypermethrin, chlorothalonil, and carbendazim, which was attributed to the enhanced activity and genes involved in pesticide metabolism [87].

Piñol and Simón [89] discussed the defensive role of BRs in protecting plants from herbicide damage. At present, the action of BRs in protecting plants from environmental stresses is clearly known, and herbicide application is considered a man-made environmental stress. Previous reports suggested that application of BRs can alleviate the decreases in plant growth caused by herbicides. Herbicides inhibit photosynthetic electron transport at PSII level compete with plastoquinone bound at the Q_B site and thus inhibit the electron transfer from Q_A to Q_B. Chlorophyll fluorescence measurements detect the effect of environmental stresses on photosynthesis. Analyses of chlorophyll fluorescence together with measures of photosynthetic CO_2 assimilation and plant growth indicated that the harmful effects caused by s-triazine herbicides can be alleviated by BRs. The protective effect of BRs persists over time and lead to a more rapid recovery of plants. BRs afford no protection against the damaging effect of other herbicides that inhibit photosynthetic electron transport but do not belong to the s-triazine group. Piñol and Simón [88] reported the protective effect of EBL on Chlorophyll (Chl) Fluorescence and Photosynthetic CO_2 Assimilation in *Vicia faba* Plants Treated with the Photosynthesis-Inhibiting Herbicide Terbutryn (Terb). Simultaneous measurements of Chl fluorescence and CO_2 assimilation in *Vicia faba* leaves were taken during the first weeks of growth to evaluate the protective effect of EBL against damage caused by the application of the herbicide terbutryn (Terb) at pre-emergence. The highest dose of Terb strongly decreased CO_2 assimilation, the maximum quantum yield of PSII photochemistry in the dark-adapted state (FV/FM), the nonphotochemical quenching (NPQ), and the effective quantum yield (DF/F0 M) during the first 3–4 weeks after plant emergence. Moreover, Terb increased the basal quantum yield of nonphotochemical processes (F0/FM), the degree of reaction center closure (1 - qp), and the fraction of light absorbed in PSII antennae that was dissipated *via* thermal energy dissipation in the antennae (1 - F0V/F0M). Application of EBL to *V. faba* seeds before sowing strongly diminished the effect of Terb on fluorescence parameters and CO_2 assimilation, which recovered 13 days after plant emergence and showed values similar to those of control plants. The protective effect of EBL on CO_2 assimilation was detected at a photosynthetic photon flux density (PFD) of 650 lmol $m^{-2}s^{-1}$ and the effect on DF/F0M and photosynthetic electron transport (J) was detected under actinic lightings up to 1750 lmol m^{-2} s^{-1}. The highest dose of EBR also counteracted the decrease in plant growth caused by Terb, and plants registered the same growth values as controls. Piñol and Simón [89] suggested that EBL could affect the Terb inhibition of PSII by displacement of Q_B from its binding site on the D_1 protein of PSII.

BRASSINOSTEROIDS AND PHYTOREMEDIATION

Barbafieri and Tassi [90] reported the possible applications of BRs in phytoremediation. Phytoremediation is a plant-based family of technologies for remediating the contamination. It has good public acceptance and is economical, compared to traditional and engineering technologies for soil treatment. Phytohormones have the specific ability to increase and support plant physiology, and are well-known and applied in horticulture, floriculture, fruit farming and other agricultural fields. The main processes of phytoremediation involve plant physiology within the plant and/or its immediate surroundings (rhizosphere); thus it can take advantage of any "assistants" that improve the efficiency of the physiological mechanisms that can make phytoremediation process more efficient. These assistants are phytohormones that have the potential to aid phytoremediation. Such treatment is harmless to the environment, practical and economically viable. Phytohormones can increase plant resistance to stress, increase plant biomass production, increase plant metal uptake, and increase organic degradation. BRs could enter into this class of phytohormone for "assisted phytoremediation by plant growth regulators". This may open a new research field, intriguing experts in both phytoremediation and phytohormones. Phytoremediation is an efficient cleanup technique based on the use of plants to remediate contamination of soil, water and air with organic or inorganic wastes. Since phytoremediation is a cost-effective, non-invasive alternative or complementary technology for transgenic-based remediation methods, thus lately it has gained attention [91]. The mechanisms of phytoremediation mainly include phytostabilization (or rhizofiltration), phytostimulation (or rhizodegradation), phytoextraction (or phytoaccumulation), phytodegradation (or phytotransformation), and phytovolatilization. Since present physical and chemical clean up technologies are very costly, this plant based remediation technique is easy and economic way to remediate the contaminated sites for mankind. Commercial phytoremediation involves about 80% organic and 20% inorganic pollutants. Phytoremediation may also become a technology of choice for remediation projects in developing countries because it is cost-efficient and easy to implement [92].

Since BRs are reported to promote seed germination, increase biomass, enhance metal uptake and stress tolerance in plants, thus they are potential candidates for phytoremediation. Exogenous application of brassinolide (BL), HBL and EBL promoted seed germination in *Eucalyptus camaldulensis*, *Lepidum sativus*, *Arachis hypogea*, *Brassica juncea*, *Oryza sativa*, *Triticum aestivum*, *Lycopersicum esculentum* and *Orabanchae minor* [59, 93-95]. BRs enhance the elongation of hypocotyls, epicotyls and peduncle of dicots, as well as coleoptiles and mesocotyls of monocots [14, 38]. Wheat grown from seed pre-treated with μM concentrations of HBL had enhanced leaf number per plant, fresh and dry weight, and activities of nitrate reductase and carbonic anhydrase [96]. Bao *et al.* [97] observed that BRs interacted synergistically with auxins to promote lateral root growth in *Arabidopsis*. BRs application promoted lateral root development by stimulating acropetal auxin transport (from base to tip) in the roots. The role of BRs in micropropagation techniques for clonal propagation of woody angiosperms was studied by Pereira-Netto *et al.* [98]. They reported that when *in vitro*-grown shoots of a hybrid between *Eucalyptus grandis and E. urophylla* were treated with 28-HCS, the treated shoots showed enhanced elongation and formation of new main shoots at low doses but there was reduced elongation and formation of primary lateral shoots. In our earlier studies, we have reported that EBL and HBL treatments (presowing) improved the shoot emergence and plant biomass production in *Brassica juncea* seedlings and plants under heavy metal stress (Cu, Zn, Mn, Co and Ni). EBL and HBL have also been found to reduce the heavy metal uptake and accumulation in *B. juncea* seedlings and plants [12, 58-61]. In rice, applications of BRs reduced the damage caused by pesticides (simazine, butachlor, or pretilachor) [62]. Further, BRs have been reported to increase the activities of antioxidant enzymes of the plants under heavy metals stress [56, 66-76].

ROLE OF BRASSINOSTEROIDS IN PLANT PRODUCTIVITY

Due to increase in human population, there is worldwide exponential increase in demand for edible crops. Therefore, it becomes necessary to increase the productivity of crops in order to sustain the ever increasing population. Chemical fertilizers have played a major role in increasing the yield of the crops but it has been established that fertilizers are one of the causes of environmental pollution posing risk to plant and human health. Hence, it becomes essential to find out environment friendly substitutes and in this regard, natural plant products are one of the emerging contenders. Khripach *et al.* [16] reported that BRs are natural, nontoxic and eco-friendly products which are applied in extremely low doses and are capable of improving the crop yield even in non-fertilized fields. BRs have been reported to influence seed germination, rhizogenesis, senescence, flowering, abscission, maturation, vegetative and reproductive developments, leaf bending, epinasty, root inhibition, induction of ethylene biosynthesis, proton-pump activation, xylem differentiation and regulation of gene expression [3, 17, 23, 62, 99, 100].

Incorporation of BRs in various field trials and agronomical practices has disclosed the fact that BRs can enhance both quality and quantity of various food crops. EBL treatment enhanced the yield of wheat, tobacco, corn, rape, orange, grape and sugar beet [74, 101, 102]. The application of HBL and EBL to potato plants in a dose of 10^{-20} mg ha^{-1} enhanced the yield by 20% and improved the quality by decreasing nitrate content and enhancing starch and vitamin C content [103]. Hnilička *et al.* [53] studied the effect of EBL on biomass and yield of wheat grain and straw grown under drought and high temperature. EBL reduced the negative effects of these stresses and enhanced the yield of wheat grain and straw. Further, Štranc *et al.* [105] observed the positive influence of BRs and lexin preparation (fulvic and humic acids mixture and auxins) on physiological state and yield of soybean. Plants treated with these preparations were found to be more resistant to short term drought, showed better physiological state and energy balance of photosynthesis and higher seed yield.

Commercially EBL is used in the form of Epin and Tianfengsu, is recommended for treatment of agricultural plants such as tomato, potato, cumber, pepper and barley to enhance their yield and quality in Russia, Belarus, Japan [106] and China. In India, Godrej Agrovet Ltd., Mumbai, has commercialized HBL for yield enhancement of grapes, groundnut, tea and peanut.

BIOSAFETY ISSUES OF BRASSINOSTEROIDS

Though BRs have not been exploited practically and comprehensively in field trials, hitherto the reports available suggest their possible impending relevance in human welfare from the practical point of view. As BRs are natural, nontoxic and eco-friendly plant products their large scale use is not expected to raise any ethical issues. Hence, it makes of BRs suitable candidates for their application in therapeutics, besides their use in agriculture. Before their commercial use in human welfare, scientists have studied their biosafety. Since BRs are natural constituents of all plants, BRs are regularly consumed by mammals. The confirmation of their safety was obtained from toxicological studies made in the Sanitary-Hygienic Institute of Belarus for EBL. Those studies revealed that the commercial formulation Epin™ (0.025 % solution of 24-epibrassinolide), in mice and rats (orally and dermally), has an LD50 of more than 5,000 mg kg^{-1} as compared to LD 50 of EBL which is more than 1000 mg kg^{-1} in mice (orally) and more than 2000 mg kg^{-1} in rats (orally and dermally). In another study, it was confirmed that 0.2 % EBL or a solution of Epin™ when applied to rabbit eyes did not cause any irritation to mucous membranes. Futhermore, the Ames test carried out at the Scientific Research Center of Toxicologic and Hygienic Regulation of Biopreparations of Russia, gave negative results for mutagenic activity. Studies on fish toxicity also showed no negative effects, but showed toxico-protective properties [16]. In micro-nuclear or chromosome aberration tests (mice CBAB1/6), neither EBL nor Epin™ caused spontaneous mutations. Biological studies carried on *Tetrahymena pyriformis* have confirmed the genetic safety of EBL and the absence of mutagenic activity over seven generations.

Recently, the developmental toxicity of HBL in wistar rats was studied at the International Institute of Biotechnology and Toxicology (IIBAT) in Tamil Nadu, India [107]. HBL was administered by oral gavage at doses of 0, 100, and 1000 mg/kg of body weight, in water, during gestation days (GD) 6 to 15, in groups of 20 mated females. Maternal and embryo-fetal toxicity was analyzed by studying effects such as clinical signs, mortality/morbidity, abortions, body weight, feed consumption, pregnancy data, gravid uterine weights, implantation losses, litter size, external, visceral, and skeletal malformations. No treatment-related effect was observed on any of the maternal/fetal end points in any dose group. Thus, it was concluded that HBL is non-teratogenic at doses as high as up to 1000 mg/kg body weight in wistar rats.

Antigenotoxic Activities of Brassinosteroids

Exogenous applications of EBL have been observed to have antigenotoxic potential. *Allium cepa* chromosomal aberration bioassay revealed the antigenotoxicity of EBL treatments [108]. Highest dose of EBL (0.5 ppm) was reported to be effective in increasing mean root lengths and lessening the number of mitoses as compared to control. However, Low doses of EBL (0.005 ppm) and intermediate doses of EBL (0.05 ppm) nearly doubled the mean root length and the number of mitosis over that of controls. Sondhi *et al.* [104] isolated and characterized the EBL from leaves of *Aegle marmelos* Correa. (Rutaceae) which was further evaluated for the antigenotoxicity against maleic hydrazide (MH) induced genotoxicity in *Allium cepa* chromosomal aberration assay. It was observed that the percentage of chromosomal aberrations induced by maleic hydrazide (0.01%) declined significantly with the treatment of EBL. Thus, EBL (10^{-7} M) proved to be the most effective concentration with 91.8% inhibition.

CONCLUSIONS

Brassinosteroids (BRs) are endogenous plant polyhydroxysteroids essential for normal growth and development of plants. Prior reports exposed the importance of BR treatment in terms of improved growth rate, yield, quality, abiotic stress tolerance and disease resistance in a variety of plants. Further, stress protective properties of BRs indicate that exogenous applications of BRs can act as immuno-modulators, if applied at low doses and at the correct stage of plant development. Being natural, nontoxic, non-hazardous, biosafe, antigenotoxic and eco-friendly plant products, BRs have been implicated in various agronomical practices. However, plant pollens are reported to be involved in biostimulation in folk medicine and form the basis for the production of some anti-inflammatory and metabolism stimulative medicines, which are especially recommended for children and elderly people with chronic infections. BRs are highly concentrated in pollens which further suggest their bright prospects in traditional folk medicines. Although it is too early to predict their clinical utility, but due to their antiviral, antifungal, antibacterial and anticancer properties, BRs are projected as potential drugs against viruses, bacteria, fungi and other pathogens. BRs have been speculated to act *via* receptor/ligand complex that binds to nuclear or cytoplasmic sites to regulate the expression of growth specific or stress protective genes. Thus, a deep understanding of the underlying molecular mechanisms of BRs action is expected to contribute to their future applications in conventional agricultural practices as well as in therapeutics for mankind.

REFERENCES

[1]   Janeczko A, Skoczowski A. Mammalian sex hormones in plants. Folia Histochem Cytobiol 2005; 43: 71-9.
[2]   Bedi NK, Virk GS, Thukral AK. Effect of steroid hormones and kinetin on mitosis in the root tips of *Allium cepa*. J Plant Sci Res 1992; 8: 43-5.
[3]   Bhardwaj R, Thukral AK. Animal hormones and plants. In: Thukral AK, Virk GS, Eds. Environmental Protection. Jodhpur: Scientific Publishers 2000; pp. 130-41.
[4]   Czerpak R, Szamrej IK. The effect of β-estradiol and corticosteroids on chlorophylls and carotenoids content in *Wolffia arrhiza* (L.) Wimm. (Lemnaceae) growing in municipal bialystok tap water. Pol J Environ Stud 2003; 12: 677-84.
[5]   Dogra R, Thukral AK. Effect of steroid and plant hormones on some germination aspects of *Triticum aestivum* L. In: Dhir KK, Dua IS, Chark KS, Eds. New Trends in Plant Physiology. New Delhi: Today and Tomorrow's Printers and Publishers 1991; pp. 65-70.
[6]   Dogra R, Thukral AK. Proteins, nucleic acids and some enzyme activities in maize plants as affected by presowing seed treatment with steroids. Indian J Plant Physiol 1994; 37: 164-8.

[7]   Szamrej IK, Czerpak R. The effect of sex steroids and corticosteroids on the content of soluble proteins, nucleic acids and reducing sugars in *Wolffia arrhiza* (L.) Wimm. (Lemnaceae). Pol J Environ Stud 2004; 13: 565-71.

[8]   Mitchell JW, Mandava NB, Worley JF, Plimmer JR, Smith MV. Brassins: a new family of plant hormones from rape pollen. Nature 1970; 225: 1065-6.

[9]   Grove MD, Spencer GF, Rohwedder WK, *et al.* Brassinolide, a plant growth promoting steroid isolated from Brassica napus pollen. Nature 1979; 281: 216-7.

[10]   Yokota T, Arima M, Takahashi N. Castasterone, a new phytosterol with planthormone potency from chestnut insect gall. Tetrahedron Lett 1982; 23: 1275-8.

[11]   Bhardwaj R, Arora HK, Nagar PK, Thukral AK. Brassinosteroids-A novel group of plant hormones. In: Trivedi PC, Ed. Plant Molecular Physiology-Current Scenario and Future Projections. Jaipur: Aaviskar Publisher 2006; pp. 58-84.

[12]   Bhardwaj R, Kaur S, Nagar PK, Arora HK. Isolation and characterization of brassinosteroids from immature seeds of *Camellia sinensis* (O) Kuntze. Plant Growth Regul 2007; 53: 1-5.

[13]   Clouse SD, Sasse JM. Brassinosteriods: essential regulators of plant growth and development. Ann Rev Plant Physiol Plant Mol Biol 1998; 49: 427-51.

[14]   Mandava NB. Plant growth promoting brassinosteroids. Ann Rev Plant Physiol Plant Mol Biol 1988; 39: 23-52.

[15]   Bajguz A, Tretyn A. The chemical characteristic and distribution of brassinosteroids in plants. Phytochem 2003; 62: 1027-46.

[16]   Khripach VA, Zhabinskii VN, de Groot A. Twenty years of Brassinosteriods: steroidal plant hormones warrant better crops for the XXI century. Ann Bot 2000; 86: 441-7.

[17]   Cerana R, Bonetti A, Marre MT, Romani G, Lado P, Marre E. Effects of a brassinosteroid on growth and electrogenic proton extrusion in azuki bean epicotyls. Physiol Plant 1983; 59: 23-7.

[18]   Kemmerling B, Schwedt A, Rodriguez P, *et al.* The BRI1-Associated Kinase 1, BAK1 has a brassinolide-independent role in plant cell-death control. Curr Biol 2007; 17: 1116-22.

[19]   Krishna P. Brassinosteriods- mediated stress responses. J Plant Growth Regul 2003; 22: 289-97.

[20]   Fariduddin Q, Yusuf M, Hayat S, Ahmad, A. Effect of 28-homobrassinolide on antioxidant capacity and photosynthesis in *Brassica juncea* plants exposed to different levels of copper. Environ Exp Bot 2009; 66: 418-24.

[21]   Xia XJ, Huang LF, Zhou YH, *et al.* Brassinosteroids promote photosynthesis and growth by enhancing activation of rubisco and expression of photosynthetic genes in *Cucumis sativus*. Planta 2009; 230: 1185.

[22]   Ali B, Hayat S, Ahmad A. Response of germinating seeds of *Cicer arietinum* to 28-homobrassinolide and/or potassium. Gen App Plant Physiol 2005; 31: 55-63.

[23]   Cao S, Xu Q, Cao Y, *et al.* Loss of function mutations in DET2 gene lead to an enhanced resistance to oxidative stress in *Arabidopsis.* Physiol Plant 2005; 123: 57-66.

[24]   Yu JQ, Huang LF, Hu WH, *et al.* A role for brassinosteriods in the regulation of photosynthesis in *Cucumis sativus*. J Exp Bot 2004; 55: 1135-43.

[25]   Bajguz A. Brassinosteroid enhanced the level of abscisic acid in *Chlorella vulgaris* subjected to short-term heat stress. J Plant Physiol 2009; 166: 882-86.

[26]   Bajguz, A. Suppression of *Chlorella vulgaris* growth by cadmium, lead, and copper stress and its restoration by endogenous brassinolide. Arch Environ Contam Toxicol 2010; 60: 406-16.

[27]   Bajguz A, Hayat S. Effects of brassinosteroids on the plant responses to environmental stresses. Plant Physiol Biochem 2009; 47: 1-8.

[28]   Divi, UK Krishna P. Brassinosteroid: a biotechnological target for enhancing crop yield and stress tolerance. New Biotech 2009; 26: 131-6.

[29]   Divi UK, Rahman T, Krishna P. Brassinosteroid-mediated stress tolerance in *Arabidopsis* shows interactions with abscisic acid, ethylene and salicylic acid pathways. BMC Plant Biol 2010; 10: 151.

[30]   Nakashita H, Yasuda M, Nitta T, *et al.* Brassinosteroids function in a broad range of disease resistance in tobacco and rice. Plant J 2003; 33: 887-98.

[31]   Ohri P, Sohal SK, Bhardwaj R, Khurma UR. Studies on the root-knot nematode, *Meloidogyne incognita*.Kofoid White) Chitwood under the influence of 24-epibrassinolide. Ann Plant Protec Sci 2008; 16:198-202.

[32]   Romanutti C, Castilla V, Cotto CE, Wachsman, MB. Antiviral effect of a synthetic brassinosteriod on the replication of vesicular stomatitis virus in vero cells. Int J Antimicrob Agents 2007; 29: 311-6.

[33]   Swaczynova J, Sisa M, Hnilickova J, Kohout L, Strnad, M. Synthesis, biological, immunological and anticancer properties of a new brassinosteriod ligand. Pol J Chem 2006; 80: 629-35.

[34]   Wachsman MB, Castilla V, Talarico LB, Ramirez JA, Galagovsky LR, Cotto CE. Antiherpetic mode of action of.22S, 23S)-3β-bromo-5α, 22, 23-trihydroxystigmastan-6-one *in vitro*. Int J Antimicrob Agents 2004; 23: 524-6.

[35]   Bancos S, Szatmari AM, Castle J, *et al*. Diurnal regulation of the brassinosteroid-biosynthetic CPD gene in *Arabidopsis*. Plant Physiol 2006; 141: 299-309.

[36]   Choi YH, Fujioka S., Nomura T, *et al*. An alternative brassinolide biosynthetic pathway *via* late C-6 oxidation. Phytochem 1997; 44: 609-13.

[37]   Fujioka S, Yokota T. Biosynthesis and metabolism of brassinosteroids. Ann Rev Plant Biol 2003; 54: 137-64.

[38]   Clouse SD. Molecular genetic studies confirm the role of brassinosteriods in plant growth and development. Plant J 1996; 10: 1-8.

[39]   Szekeres M, Nameth K, Kalmar ZK, Mathur J, Kauscrman A, Altmann T, Redei GP, Nagy F, Scrall J, Koncz C. Brassinosteriods rescue the deficiency of CYP 90, a cytochrome P450, controlling cell elongation and de-etiolation in *Arabidopsis*. Cell 1996; 85: 171-82.

[40]   Kim TW, Wang, ZY. Brassinosteroid signal transduction from receptor kinases to transcription factors. Ann Rev Plant Biol 2010; 61: 681-704.

[41]   Mittler R. Abiotic stress, the field environment and stress combination. Trends Plant Sci 2006; 11: 15-9.

[42]   Vasyukova NI, Chalenko GI, Kaneva IM, Khripach VA, Ozeretskovskaya, OL. Brassinosteroids and potato late blight. Prik Biok Mikrobiol 1994; 30: 464-70.

[43]   Wachsman MB, Lopez EM, Ramirez JA, Galagovsky LR, Cotto CE. Antiviral effect of brassinosteriod against herpes virus and arena virus. Antiviral Chem Chemother 2000; 11: 71-7.

[44]   Zullo MAT, Adam G. Brassinosteroids phytohormones–structure bioactivity and applications. Braz J Plant Physiol 2002; 14: 143-81.

[45]   Ohri P, Sohal SK, Bhardwaj R, Khurma UR. Morphogenetic and biochemical responses of the root-knot nematode, *Meloidogyne incognita* (Kofoid White) Chitwood to isolated brassinosteroids. Ann Plant Protec Sci 2007; 15: 226-31.

[46]   Richter K, Koolman, J. Antiecdysteroid effects of brassinosteriods in insects. In: Cutler HG, Yokota T, Adams G, Eds. Brassinosteriods: Chemistry, bioactivity, and application. Washington: American Chemical Society 1991; pp. 265-78.

[47]   Dinan L, Whitting P, Girault JP, *et al*. Cucurbitacins are insect steroid hormone antagonists acting at the ecdysteroid receptors. Biochem J 1997; 327: 643-50.

[48]   Robertson CW. The metamorphosis of *Drosophila melanogaster*, including an accurately timed account of the principal morphological changes. J Morphol 1936; 59: 351-99.

[49]   Jiang C, Baehrecke EH, Thummel CS. Steroid regulated programmed cell death during *Drosophila* metamorphosis. Development 1997; 124: 4673-83.

[50]   Jochova J, Zakeri Z, Lockshin RA. Rearrangement of the tubulin and actin cytoskeleton during programmed cell death in *Drosophila* salivary glands. Cell Death Diff 1997; 4: 140-9.

[51]   Lee CY, Baehrecke EH. Steroid regulation of autophagic programmed cell death during development. Development 2001; 128: 1443-55.

[52]   Roberts K, McCann, MC. Xylogenesis: The birth of a corpse. Curr Opin Plant Biol 2000 3: 517-22.

[53]   Hnilička F, Hniličková H, Martinková, JB. The influence of drought and the application of 24-epibrassinolide on the formation of dry matter and yield in wheat. Obervellach, Austria: Proc VI. Alps-Adria Scientific Workshop 2007; pp. 457-60.

[54]   Jager CE, Symons GM, Ross JJ, Reid JB. Do brassinosteroids mediate the water stress response? Physiol Plant 2008; 133: 417-25.

[55]   Sasse JM. Physiological actions of brassinosteroids. In: Sakurai A, Yokota T, Clouse SD, Eds. Brassinosteroids: Steroidal plant hormones. Tokyo: Springer-Verlag 1999; pp. 137-61.

[56]   Sharma I, Pati PK, Bhardwaj R. Regulation of growth and antioxidant enzyme activities by 28-homobrassinolide in seedlings of *Raphanus sativus* L. under cadmium stress. Indian J Biochem Biophys 2010; 47: 172-7.

[57]   Xia XJ, Huang YY, Wang L, *et al*. Pesticides-induced depression of photosynthesis was alleviated by 24-epibrassinolide pretreatment in *Cucumis sativus* L. Pest Biochem Physiol 2006; 86: 42-8.

[58]   Bhardwaj, R, Sharma, P, Arora, HK, Arora, N. 28- Homobrassinolide regulated Mn- uptake and growth of *Brassica juncea* L. Can J Pure Appl Sci 2008; 2: 149-54.

[59]   Sharma P, Bhardwaj, R. Effect of 24-epibrassinolide on seed germination, seedling groswth and heavy metal uptake in *Brassica juncea* L. Gen App Plant Physiol 2007; 33: 59-73.

[60]   Sharma P, Bhardwaj R, Arora N, Arora, HK. Effect of 28-homobrassinolide on growth, zinc metal uptake and antioxidative enzyme activities in *Brassica juncea* L. seedlings. Braz J Plant Physiol 2007; 19: 203-10.

[61]   Sharma P, Bhardwaj R, Arora N, Arora HK, Kumar A. Effects of 28-homobrassinolide on nickel uptake, protein content and antioxidative defence system in *Brassica juncea*. Biol Plant 2008; 52: 767-70.

[62]   Sasse JM. Physiological Action of Brassinosteriods: An update. J Plant Growth Regul 2003; 22: 276-88.

[63]   Almeida JM, Fidalgo F, Confraria A, Santos A, Pires H, Santos, I. Effect of hydrogen peroxide on catalase gene expression, isoform activities and levels in leaves of potato sprayed with homobrassinolide and ultrastructural changes in mesophyll cells. Function Plant Biol 2005; 32: 707-20.

[64]   Mazorra LM, Nunez M, Hechavarria M, Coll F, Sanchez-Blanco MJ. Influence of brassinosteroids on antioxidant enzymes activity in tomato under different temperatures. Biol Plant 2002; 45: 593-6.

[65]   Ozdemir F, Bor M, Demiral T, Turkan I. Effects of 24-epibrassinolide on seed germination, seedling growth, lipid peroxidation, proline content and antioxidative system of rice (*Oryza sativa* L.) under salinity stress. Plant Growth Regul 2004; 42: 203-11.

[66]   Alam MM, Hayat S, Ali B, Ahmad A. Effect of 28-homobrassinolide treatment on nickel toxicity in *Brassica juncea*. Photosynthetica 2007; 45: 139-42.

[67]   Arora N, Bhardwaj R, Sharma P, Arora HK, Arora N. Amelioration of Zn toxicity by 28-homobrassinolide in *Zea mays* L. Can J Pure App Sci 2008; 2: 503-9.

[68]   Arora N, Bhardwaj R, Sharma P, Arora HK. Effects of 28-homobrassinolide on growth, lipid peroxidation and antioxidative enzyme activities in seedlings of *Zea mays* L. under salinity stress. Acta Physiol Plant 2008; 30: 833-9.

[69]   Bhardwaj R, Sharma I, Arora N, *et al.* Prospects of brassinosteroids in medicinal applications. In: Hayat S, Brassinosteroids: A class of plant hormone. The Netherlands: Springer 2010; pp. 439-58.

[70]   Bhardwaj R, Sharma I, Arora N, *et al.* Regulation of oxidative stress by BRs in plants. In: Ahmad P, Ed. Oxidative stress: Role of antioxidants in plants. New Delhi: Stadium Press Pvt. Ltd. 2010; pp. 215-31.

[71]   Choudhary SP, Bhardwaj R, Gupta BD, Dutt P, Kanwar M, Arora, N. Epibrassinolide regulated synthesis of polyamines and auxins in Raphanus sativus L. seedlings under Cu metal stress. Braz J Plant Physiol 2009; 21: 25-32.

[72]   Choudhary SP, Bhardwaj R, Gupta BD, Dutt P, Arora N. Effect of 24-epibrassinolide on polyamines titers, antioxidative enzymes and seedling growth of Raphnus sativus L. under copper stress. Plant Stress 2009; 3: 7-12.

[73]   Hayat S, Ali B, Hassan SA, Ahmad A. Brassinosteriods enhanced antioxidants under cadmium stress in *Brassica juncea*. Environ Exp Bot 2007; 60: 33-41.

[74]   Ikekawa N, Zhao YJ. Application of 24-epibrassinolide in agriculture. In: Cutler HG, Yokota T, Adam G. Eds. Brassinosteroids Chemistry, bioactivity, and applications, ACS Symposium Series. Washington: American Chemical Society 1991; pp. 280-91.

[75]   Sirhindi G, Kumar S, Bhardwaj R, Kumar, M. Effects of 24-epibrassinolide and 28-homobrassinolide on the growth and antioxidant enzyme activities in the seedlings of *Brassica juncea* L. Physiol Mol Biol Plants 2009; 15: 335-41.

[76]   Pal S, Bhardwaj R, Gupta BD, *et al.* Epibrassinolide induces changes in indole-3-acetic acid, abcisic acid and polyamine concentrations and enhances antioxidant potential of radish seedlings under copper stress. Physiol Plant 2010; 140: 280-96.

[77]   Malìková J, Swaczynová J, Kolář Z, Strnad, M. Anticancer and antiproliferative activity of natural brassinosteroids. Phytochem 2008; 69: 418-26.

[78]   Wachsman MB, Ramirez JA, Talarico LB, Galagovsky LR, Cotto CE. Antiviral activity of natural and synthetic brassinosteriods. Curr Med Chem 2004; 3: 163-79.

[79]   Wachsman MB, Ramirez JA, Galagovsky LR, Cotto CE. Antiviral activity of brassinosteriods derivatives against measles virus in cell cultures. Antiviral Chem Chemother 2002; 13: 61-6.

[80]   Michelini FM, Ramirez JA, Berra A, Galagovsky LR, Alché, LE. *In vitro* and *in vivo* antiherpetic activity of three new synthetic brassinosteroids analogues. Steroids 2004; 69: 713-20.

[81]   Khripach VA, Altsivanovich K, Zhabinskii VN, Samusevich, M. Natural plant compound with anti-HIV activity. US Patent 20040253289, 2004.

[82]   Michelini FM, Berra A, Alché LE. The *in vitro* immunomodulatory activity of a synthetic brassinosteroid analogue would account for the improvement of herpetic stromal keratitis in mice. J Ster Biochem Mol Biol 2008; 108: 164-70.

[83]   Franek F, Eckschlager T, Kohout L. 24-epibrassinolide at subnanomolar concentrations modulates growth and production characteristics of a mouse hybridoma. Collec Czechoslovak Chem Commun 2003; 68: 2190-200.

[84]   Oklestková J, Hoffmannová L, Steigerová J, Kohout L, Kolár Z, Strnad M. Natural brassinosteroids for use for treating hyperproliferation, treating proliferative diseases and reducing adverse effects of steroid dysfunction in mammals, pharmaceutical composition and its use. US Patent 20100204460, 2010.

[85]   Mäenpää KA, Sormunen AJ, Kukkonen JVK. Bioaccumulation and toxicity of sediment associated herbicides (ioxynil, pendimethalin, and bentazone) in *Lumbriculus variegatus*.Oligochaeta) and *Chironomus riparius* (Insecta). Ecotoxicol Environ Safety 2003; 56: 398-410.

[86]   Hamada K. Brassinolide in crop cultivation. In: Macgregor P, Ed. Plant growth regulators in agriculture. Taiwan: Food Fertility Technology, Central Asia Pacific Region 1986; pp. 190-196.

[87]   Xia XJ, Zhang Y, Wu JX, Wang JT, Zhou YH, Shi K, Yu YL, Yu JQ. Brassinosteroids promote metabolism of pesticides in cucumber. J Agricult Food Chem 2009; 57: 8406-13.

[88]   Piñol R, Simón E. Effect of 24-epibrassinolide on chlorophyll fluorescence and photosynthetic CO_2 assimilation in *Vicia faba* plants treated with the photosynthesis-inhibiting herbicide terbutryn. J Plant Growth Regul 2009; 28: 97-105.

[89]   Piñol R, Simón E. Protective effects of brassinosteroids against herbicides. In: Hayat S, Ahmed A, Eds. Brassinosteroids: A Class of Plant Hormone. Heidelberg, London, New York, Dordrecht: Springer 2011; pp. 309-44.

[90]   Barbafieri M, Tassi E. Brassinosteroids for phytoremediation application. In: Hayat S, Ahmed A, Eds. Brassinosteroids: A Class of Plant Hormone. Heidelberg, London, New York, Dordrecht: Springer 2011; pp. 403-37.

[91]   Pilon-Smits E. Phytoremediation. Ann Rev Plant Biol 2005; 56: 15-39.

[92]   Sarma H. Metal hyperaccumulation in plants: a review focusing on phytoremediation technology. J Environ Sci Technol 2011; 4: 118-38.

[93]   Rao SSR, Vardhini BV, Sujatha E, Anuradha S. Brassinosteriods-A new class of phytohormones. Curr Sci 2002; 82: 1239-45.

[94]   Sasse JM, Smith R, Hudson I. Effect of 24-epibrassinolide on germination of seeds of *Eucalyptus camaldulensis* in saline conditions. Proc Plant Growth Regul Soc Am 1995; 22: 136-41.

[95]   Vardhini BV, Rao SSR. Effect of brassinosteriod on growth, metabolite content and yield of *Arachis hypogea*. Phytochem 1998; 48: 927-30.

[96]   Hayat S, Ahmad A, Hussain A, Mobin M. Growth of wheat seedlings raised from the grains treated with 28-homobrassinolide. Acta Physiol Plant 2001; 23: 27-30.

[97]   Bao F, Shen J, Brady SR, Muday GK, Asami T, Yang Z. Brassinosteroids interact with auxin to promote lateral root development in *Arabidopsis*. Plant Physiol 2004; 134: 1624-31.

[98]   Pereira-Netto AB, Carvalho-Oliveira MMC, Ramirez JA, Galagovsky LR. Shooting control in *Eucalyptus grandis* X *E. urophylla* hybrid: comparative effects of 28-homocastasterone and a 5 α-monofluoro derivative. Plant Cell Tiss Org Cult 2006; 86: 329-35.

[99]   Çağ S, Gören-Sağlam N, Çingil-Bariş Ç, Kaplan E. The effect of different concentration of epibrassinolide on chlorophyll, protein and anthocyanin content and peroxidase activity in excised red cabbage (*Brassica oleracea* L.) cotyledons. Biotechnol Biotechnol Equip 2007; 21: 422-5.

[100]   Leubner-Metzger G. Brassinosteroids and gibberellins promote tobacco seed germination by distinct pathways. Planta 2001; 213: 758-63.

[101]   Hu AS, Jiang BF, Guan YL, Mou H. Effects of epi-brassinolide on abscission of young fruit explants and cellulose activity in abscission zone of citrus. Plant Physiol Commun 1990; 5: 24-6.

[102]   Zhao YJ, Chen J. Studies on physiological action and application of 24-epibrassinolide in agriculture. In: Hayat S, Ahmad A, Eds. Brassinosteroids: bioactivity and crop productivity, The Netherland: Kluwer Academic Publishers 2003; pp. 159-70.

[103]   Khripach VA, Voronina LV, Malevannaya NN. Preparation for the diminishing of heavy metals accumulation by agricultural plants. RU Patent Applied 95, 101, 850, 1996.

[104]   Sondhi N, Bhardwaj R, Kaur S, Kumar N, Singh B. Isolation of 24-epibrassinolide from leaves of Aegle marmelos and evaluation of its antigenotoxicity employing *Allium cepa* chromosomal aberration assay. Plant Growth Regul 2008; 54: 217-24.

[105]   Stranc P, Becka D, Vasak J, Stranc J, Stranc D. The effect of protective seed mixture on damage of stems of winter oilseed rape (*Brassica napus* L.) by rapeseed stem weevil (*Ceutorhynchus napi*) and cabbage stem weevil (*Ceutorhynchus pallidactylus*). Sci Agricult Bohemica 2008; 39: 16-23.

[106]   Moiseev A. Preparation epin: Water of life. Eureka 1998, 68.

[107] Murkunde YV, Murthy PB. Developmental toxicity of homobrassinolide in wistar rats. Int J Toxicol 2010; 29: 517-22.

[108] Howell WM, Keller GE, Kirkpatrick JD, Jenkins RL, Hunsinger RN, McLaughlin, EW. Effects of the plant steroidal hormone, 24-epibrassinolide, on the mitotic index and growth of onion (*Allium cepa*) root tips. Genet Mol Res 2007; 6: 50-8.

CHAPTER 2

Mitigation of Water Stress and Saline Stress by Brassinosteroids

B. Vidya Vardhini[*]

Department of Botany, Telangana University, Dichpally, Nizamabad -503322, Andhra Pradesh, India

Abstract: Brassinosteroids is a new group of plant growth regulators with significant growth promoting influence which was first isolated and characterized from the pollen of *Brassica napus* L. Subsequently, they were reported from monocots, dicots, gymnosperms, pteridophyte and alga. Brassinosteroids are now considered as plant hormones with pleiotropic effects as they influence varied developmental processes like growth, germination of seeds, rhizogenesis, flowering, senescence and all kinds of stresses. This chapter reviews the effect of brassinosteroids in mitigating water and saline stresses.

Keywords: 24-epibrassinolide, abiotic stresses, drought resistance, drought stress, homobrassinolide, osmotic stress.

INTRODUCTION

Brassinosteroids (BRs) are a new type of polyhydroxylated steroid phytohormones with significant growth promoting activity [1-4]. Brassinolide (BL), the first BR, was discovered in 1970 by Mitchell and his co-workers [5] and later extracted from the pollen of *Brassica napus* L [6]. BRs are considered ubiquitous in plant kingdom as they are found in almost all the phyla of the plant kingdom like alga, pteridophyte, gymnosperms, dicots and monocots [7]. BRs satisfy all the pre-requisites of typical plant growth regulators, *i.e.*, natural in occurrence, mobile, and active at extremely low concentrations [8]. Up till now approximately 60 BR-related compounds have been identified [9].

The work with BR biosynthetic mutants in *Arabidopsis thaliana* [10] and *Pisum sativum* [11] have provided strong evidences that BRs are essential for plant growth and development. A simple BR- analogue 2α, 3α-dihydroxy-17β-(3-methyl butynyloxyl) 7-oxa-B-homo-5α androstan-6-one induces bean second node splitting which is considered as the prominent physiological feature of BRs [12]. Dwarf and de-etiolated phenotypes and BR – deficient species of some *Arabidopsis* mutants were rescued by application of BRs [13, 14]. Even *Pharbitis nil Uzukobito*, a defective BR- biosynthetic dwarf mutant was rescued by application of BRs [15] which emphasized that BR-deficient and defective BR-biosynthetic species exhibited abnormal growth. Friedrichsen *et al.* [16] reported that three redundant BR genes encode transcription factors which are required for normal growth, indicating the necessity of BRs for normal growth. Similarly, the inhibition of growth [17] and secondary xylem development [18] of cress (*Lepidius sativus*) by brassinozole, a specific inhibitor of BL synthesis was reversed by the exogenous application of BL, also indicating the necessity of BRs for normal plant growth. Further, Shimada *et al.* [19] also reported that BR-biosynthetic genes exhibited organ specific expression in *Arabidopsis thaliana*.

Arteca and Arteca [20] reported that BRs induce exaggerated growth in hydroponically grown *Arabidopsis thaliana*, besides controlling the proliferation of leaf cells [21]. In addition, BRs promote growth of apical meristems in potato tubers [22], accelerate the rate of cell division in isolated protoplasts of *Petunia 16ynthe* [23], and induce callus growth and regeneration ability in *Spartina patens* (Poaceae) [24]. BRs play a prominent role in nodulation and nitrogenase activity in groundnut [25], 16ynthe bean [26] and soya bean [27]. BRs also accelerate ripening of tomato pericarp discs [28] and litchi pericarps [29], besides playing a pivotal role in the regulation of expression of key genes involved in *Arabidopsis thaliana* anther and pollen development [30]. BRs exhibit synergistic relationships with other plant growth regulators like indol acetic

*Address correspondence to B. Vidya Vardhini: Department of Botany, Telangana University, Dichpally, Nizamabad -503322, Andhra Pradesh, India; E-mail: drvidyavardhini@rediffmail.com

acid and benzyladenine in the secondary metabolism of cultured *Onosma paniculatum* cells [31], ethylene production and ABA content in potato tubers [22], maize [32] and in many other crop plants [33]. Saka *et al.* [34] stated that BL application at the meiosis and flowering stages affected the levels of endogenous plant hormones during the grain filling stages. The role of BRs in ethylene production along with auxins and cytokinins in *Arabidopsis thaliana* has been recently reported [35].

Although BRs were initially identified based on their growth promoting activities, subsequent physiological and genetic studies revealed additional functions of BRs in regulating a wide range of processes, including source/sink relationships, seed germination, photosynthesis, senescence, photomorphogenesis, flowering and responses to different abiotic and biotic stresses [36, 37]. Enhancement of resistance of plants to various environmental stresses by BRs has been evaluated aiming at finding practical applications for BRs in agriculture [38]. Even Sasse [39] reported that BRs interact with environmental signals such as light, growth and temperature and stimulate the synthesis of particular proteins. Du and Pooviah [40] also reported that BRs are plant-specific steroid hormones that have an important role in coupling environmental factors, especially light, with plant growth and development. However, those authors stated that the role of changes in the endogenous BRs on the response(s) to the environmental stimuli was still largely unknown.

Various data provided consistent evidence that exogenous BRs are effective in stressful rather than optimal conditions [41]. However, it is not yet confirmed whether endogenous levels of BRs in stressed and unstressed plants reflect the differential response between stressful and optimal conditions [42, 43]. But, the molecular analysis of *Arabidopsis* mutants displaying 17synthesis17 elongation defects in both dark and light revealed that deficiencies in BR-biosynthesis and signaling permit de-repression of genes induced by stress in light [44]. BRs have been further explored for stress-protective properties in plants against a number of stresses like chilling [45, 46], salt [47], heat [48], water [49] and heavy metals [50, 51] and oxidative [52] stresses, as well. Furthermore, Haubrick and Assmann [9] have pointed out that BRs regulate various physiological processes during various stresses. Recently, Xia *et al.* [53] aptly stated that BRs induce plant tolerance to a wide spectrum of stresses. This chapter reviews the effects of BRs on the mitigation of saline and water stresses. The mitigative role of BRs on plants under drought/water stress and salt/saline stress are presented as:

1. Seedling Growth

2. Plant Growth and Yield

3. Photosynthesis

4. Metabolites and Enzymes

5. Other aspects

MITIGATION OF DROUGHT/WATER/STRESS BY BRS

In plants, one of the most promising effects of BRs, apart from growth stimulation, is their ability to confer resistance against various abiotic stresses. Based on application studies, exogenously applied bioactive BRs have been shown to improve various aspects of plant growth under water stress conditions. Earlier, Shen *et al.* [54] studied the physiological effects of BL on drought resistance in maize plants. BL, at a concentration of 12.5 mg^{1-1}, increased the percentage of survival of callus tissues of a drought resistant variety (PAN 6043), and a drought sensitive variety (SC 701) of maize [55, 56]. DAA-6, a BR- analogue, exhibited promotive effect on the development of sugarcane plantlets regenerated from callus subjected to water stress [57]. Kagale *et al.* [58] reported that BRs confer tolerance to a wide range of stresses in *Arabidopsis thaliana* and *Brassica napus*. The studies conducted by Divi *et al.* [59] revealed that BRs mediate the stress tolerance in *Arabidopsis* by interacting with abscisic acid, ethylene and salicylic acid. However, studies conducted by Jager *et al.* [60] on BR mutants *lkb* (a BR-deficient mutant) and *lka* (a BR-perception mutant) of pea, and on wild type plants showed that endogenous BR levels are not normally a part of the plants's responses to water stress.

Seedling Growth

Sairam *et al.* [61] observed that treatment of wheat seeds with 28-homobrassinolide (28-homoBL) resulted in increased germination, amylase activity and total proteins in 48 days old seedlings, and increased shoot length in 96 days old seedlings, with or without PEG (polyethylene glycol)-6000 induced-water stress. In maize, BRs have been shown to mitigate the negative effect of water stress *via* the antioxidative system in both, drought sensitive and drought resistant cultivars [62]. Rajasekaran and Blake [63] reported that BRs protect the photosynthesis and enhance the growth of jack pine seedlings under water stress. BRs were also found to alleviate the negative effect of water stress in sorghum seedlings by enhancing the rate of seed germination, as well as seedling growth [64]. Further, BRs were also found to ameliorate the growth inhibitory effect of PEG-induced osmotic stress in seed germination and seedling growth of sorghum [65], where this amelioration resulted in enhanced metabolite contents like soluble proteins and free proline. Pot experiments designed to investigate the effects of BL on 1-year-old *Robinia pseudoacacia* L. seedlings grown under moderate or severe water stress, compared to untreated seedlings, showed that treatment with 0.2 mg/l BL decreased the transpiration rate, stomatal conductance and malondialdehyde (MDA) content, indicating that application of BL can ameliorate the effects of water stress and enhance drought resistance in *Robinia* seedlings [66]. The supplementation of BL resulted in decreased stomatal conductance atherby resulting in reduced transpiration rates. Yuan *et al.* [67] reported that BR treatment in tomato (*Lycopersicon esculentum*) seedlings under water stress resulted in enhanced stomatal conductance and net photosynthetic rate suggesting that the amelioration of the drought stress of tomato seedlings may be due to epiBL-induced elevation of endogenous ABA. EpiBL-application enhanced the net CO_2 intake for the efficient photosynthesis by enhancing the stomatal conductance and further emphasized that the elevation of ABA resulted in increased antioxidant enzymes which puts a new insight on BR-research.

Plant Growth and Yield

Schilling *et al.* [68] examined the effects of homoBL on sugar beet under drought stress and found an increase of the tap root mass, sucrose content and sucrose yield in plants under drought stress. In two varieties of wheat, Sairam [69] observed enhanced grain yield and yield related parameters under both, irrigated and moisture stress conditions, after treatment with 28-homoBL. In a related study, the increase in seed yield after application of homoBL in wheat plants under water stress was associated with increases in ear number per plant, grain number per ear, 1,000 grains weight and harvest index [61]. Treatment with 24-epibrassinolide (24-epiBL) enhanced growth of juvenile gram (*Cicer arientium*) plants subjected to water stress conditions [70]. Foliar application of 0.4 ppm BRs at pre-flowering stage and pod development stage increased the oil yield of mustard (*Brassica juncea*) under water deficit conditions [71]. Winter wheat varieties *Ebi, Estica, Samanta* grown under drought stress conditions supplemented with 24-epiBL were found to show reduced negative effect of the monitored drought stress in all the three varieties and increased dry matter contents, as well as yields [72]. 24-EpiBL also played a pivotal role even in alga by inducing increased growth and biomass content of *Spirulina platensis*, a *Cyanophyceae*, grown under NaCl stress [73].

Photosynthesis

In two wheat varieties, Sairam [69] observed enhanced chlorophyll content and photosynthesis under irrigated and water deficit conditions after treatment with 28-homoBL. Pre-treatment of seeds with epiBL resulted in marked increases in the rate of photosynthesis in spring wheat under drought stress [74]. Exogenously applied 24-homoBL increased the photosynthetic rate in Indian mustard plants under drought stress [75]. 24-EpiBL also enhanced the photosynthetic capacity of cauliflower plants [76] subjected to water stress. BL alleviated the adverse effects of water deficits in soybean by increasing the photosynthetic rate and the antioxidative system [77].

Metabolites and Enzymes

Pustovaitova *et al.* [78] reported that epiBL increased drought resistance of cucumber plants by enhancing the contents of free amino acids and amides. Application of 28-HomoBL to wheat plants has been shown to enhance soluble protein content, relative water content, nitrate reductase activity, chlorophyll content and photosynthesis in plants subjected to water deficit Sairam [69]. Recent studies by Behnamnia *et al.* [79] on the effects of 24-epiBL on tomato plants grown under water stress revealed that BRs considerably alleviated the oxidative damage that occurred under drought stress by increasing the activities of the antioxidant enzymes peroxidase (POD), catalase (CAT), ascorbate peroxidase (APX), glutathione reductase (GR), and superoxide dismutase (SOD)].

Other Aspects

Schlling *et al.* [69] reported that treatment of sugar beet subjected to water stress with homoBL enhanced taproot mass, acid invertases activity and sucrose content, as well sucrose synthetase activity. Upreti *et al.* [26] reported that BRs (epiBL and homoBL) enhance root nodulation and phytochrome content in French bean submitted to water deficit. Farooq *et al.* [80] reported that exogenously applied 28-homoBL and 24-epiBL, both used as seed priming and foliar spray, produced profound changes that improved drought tolerance in fine grain aromatic rice by improving the net CO_2 assimilation, water use efficiency, leaf water status, membrane properties and increased production of free proline. Treatmet of sorghum plants with 24-epiBL resulted in increased relative water content and decreased stomatal transpiration rate [81, 82].

MITIGATION OF SALINITY /SALT STRESS BY BRs

BRs increase the resistance of plants to a variety of stresses, including salinity stress. Hathout [83] demonstrated the counter action of BL on salinity stress in wheat plants. The ability of BL, 24-epiBL and 28-homoBL to counter act the salinity stress-induced growth inhibition of ground nut (*Arachis hypogaea* L) was reported by Vardhini and Rao [84]. Improved salt tolerance after BR treatment was confirmed in rice crops [85]. Ali *et al.* [86] demonstrate that 24-epiBL protected *Brassica junceae* against the stress generated by salinity and nickel. Even the endogenous BRs are positively involved in the plant's responses to salt stress in BR-deficient *Arabidopsis* mutants like *det-2-1* and *bin2-1*, in which exogenous BRs clearly improved salt tolerance and increased growth and photosynthetic activity [87]. Treatment of *Hordeum vulgare* and *Phaseolus vulgaris* seeds with BRs enhanced salt tolerance for both plants [88]. 28-HomoBL also reduced the negative impact of salinity when applied as seed treatment to *Brassica junceae* plants [89]. EpiBL proved to be an antagonist of saline stress in salt-treated roots of *Vigna 19ynthes* [90]. In addition, Amzallang [91] claimed that BR (epiBL) was a modulator of salt stress adaption in *Sorghum bicolour*. However, the study conducted by Qayyam *et al.* [92] did not give a promising result for the use of foliar application of 24-epiBL on wheat plants grown under saline conditions, once growth and photosynthetic characteristics were not significantly changed after 24-epiBL.

Seedling Growth

In the case of *Eucalyptus camuldensis*, treatment of seeds with 24-epiBL resulted in increase in seed germination under saline conditions of 50 mM NaCl [93]. Similarly, 24-epiBL and 28-homoBL were found to alleviate the salinity-induced inhibition of germination and seedling growth in rice [94], and also to improve the photosynthetic pigment levels and nitrate reductase activity [95]. EpiBL reduced the salinity-induced accumulation of abscisic acid and wheat germ agglutinin, besides partially restoring growth in the 4 day old roots of wheat by increasing accumulation of lectin, probably indicating the ABA-independent control of lectin content by epiBL [96]. The effects of 24-epiBL on seedling growth, antioxidative system, lipid peroxidation, proline and soluble protein content, investigated in seedlings of the salt-sensitive rice cultivar IR-28, clearly demonstrated that 24-epiBL treatment considerably alleviated oxidative damage that occurred under NaCl-stressed conditions and improved seedling growth [47]. Analysis of the effects of 28-homoBL and 24-epiBL on growth and on the activity of antioxidant enzymes in rice seedlings grown under salinity stress demonstrated that the ameliorating ability of BRs is related to increased scavenging of reactive oxygen species (ROS), which reduces the oxidative stress induced by NaCl and stimulate growth of rice seedlings under saline stress [97]. 28-HomoBL nullified the negative impact of saline stress in seedlings of *Zea mays* by enhancing seedling growth, lipid peroxidation and antioxidative system [98]. 24-EpiBL also increased the rate of seed germination and seedling growth of barley [99]. Finally, Divi *et al.* [59] demonstrated that salt stress-induced inhibition of seed germination of *Arabidopsis thaliana* mutants ein2 (sensitive to ethylene) and aba1-1 (ABA deficient mutant) were rescued by the exogenous supplementation with BRs.

Plant Growth and Yield

BL was found to overcome the inhibitory effect of salt stress and to improve growth of rice plants [100]. Similarly, treatment of barley seed with epiBL and homoBL removed the inhibitory effect of salt stress on the growth of barley crops [101, 102]. Exogenous application of BRs, as foliar spray, was found to increase growth of two hexaploid wheat (*Triticum aestivum* L.) cultivars viz., S-24 (salt tolerant) and MH-97 (moderately salt sensitive) subjected to saline stress [103, 104]. 24-EpiBL ameliorated the inhibitory effect of saline stress in *Capsicum annuum* cv. Beldi (pepper) plants and enhanced shoot growth, biomass, besides protecting the integrity of the cellular membrane in salt stressed plants [105]. *Brassica junceae* plants subjected to saline stress exhibited the ability to overcome the negative impact of NaCl when treated with 28-homoBL [106].

Photosynthesis

Improvement in growth of two wheat cultivars viz., S-24 (salt tolerant) and MH-97 (moderately salt sensitive) after foliar application of 24-epiBL was found to be associated with enhanced photosynthetic capacity of wheat plants [104]. 24-EpiBL has been shown to have a protective effect on cells ultra structure in leaves placed under saline stress and further prevented nuclei and chloroplast degradation, paving a way for better photosynthesis [107]. 28-homoBL has also been shown to enhance the photosynthetic efficiency of the chick pea [108] and *Brassica junceae* [109] subjected to saline stress. The root – application of 24-epiBL was also able to enhance the photosynthetic capacity of wheat plants [110] under salt stressed conditions. Further, 24-epiBL supplied as foliar treatment also enhanced the photosynthetic parameters of *Brassica junceae* under salt stressed conditions [86].

Metabolites and Enzymes

28-HomoBL was found to alleviate the oxidative stress in salt-treated maize plants by acting *via* the oxidative enzymes like SOD, CAT, GR, APX and guaicol peroxidase [98]. A BR- analogue, polyhydroxylated spirostanic (BB-16), applied to rice seedlings grown *in vitro* in culture medium supplemented with NaCl induced significant increase in the activities of CAT, SOD and GR, and a slight increase in APX, indicating its ability to confer tolerance to seedlings against saline stress [111]. 24-EpiBL was found to increase the free proline content of *Spirulina platensis*, a cyanophyceae, grown under NaCl stress [73]. Application of BRs to seeds of *Phaseolus vulgaris and Hordeum vulgare* increased their salt tolerance [88]. Further, BRs were also found to increase the levels of osmolytes like proline, glutathione and betain in *Hordeum vulgare* and *Phaseolus vulgaris* crops subjected to salt stress, indicating their halotrophic ability [112].

Other Aspects

EpiBL – 55 mitigated the saline stress in barley root tip cells by stimulating mitotic activity and also by reducing the frequency of chromosome aberrations [113, 114]. Hayat *et al.* [115] reported that 28-homoBL increased the levels of the plant hormone ethylene in mustard plants subjected to saline stress. Application of BL inhibited the light-dependent induction of proline biosynthesis by abscisic acid, a plant growth retardant, in *Arabidopsis* grown under saline conditions [116]. Further, 24-EpiBL also alleviated the negative effect of sodium chloride in wheat plants by increasing cytokinins content and decreasing abscisic acid content [117].

Several studies clearly revealed that BR application to crop plants helps them to overcome environmental stresses [118, 119] and also to increase crop yields through their growth promoting and anti stress effects [120]. The molecular analysis of *Arabidopsis* mutants displaying 20 synthesis 20 elongation defects in both dark and light revealed that deficiencies in BR biosynthesis and signaling permit de-repression of stress induced genes in light [44]. Further progress in the analysis of *Arabidopsis thaliana* mutants defective in BR biosynthesis or signaling is expected to provide a deeper insight into the evolution of steroid hormone regulation in eukaryotes, as well as into the mechanisms by which BRs control basic functions such as cell elongation, morphogenesis and stress responses [121].

Kumaro and Takatsuto [122], who were impressed by the ability of BRs to induce plant resistance against various environmental stresses, stated that "*the role of brassinosteroids in protecting plants against environmental stresses will be an important research theme for clarifying the mode of action of brassinosteroids and may contribute greatly to the usage of brassinosteroids in agricultural production*". Lately, Li *et al.* [66] also emphatically stated that "*Treatment of seedlings with BL may be a useful management tool for afforestation projects in arid and semiarid areas*".

ACKNOWLEDGEMENTS

The author thanks Prof. S. Seeta Ram Rao, Department of Botany, Osmania University, for his critical suggestions in this chapter. The financial support to Dr. B. Vidya Vardhini, from University Grants Commission (UGC), New Delhi, India is gratefully acknowledged.

REFERENCES

[1]   Tanak K, Asami T, Yoshida S, Nakamura Y, Matsuo T, Okamat S. Brassinosteroid homeostatis in *Arabidopsis* is ensured by feedback expressions of multiple genes involved in its metabolism. Plant Physiol 2005: 138; 1117-25.

[2]   Vardhini BV, Anuradha S, Rao SSR. Brassinosteroids-New class of plant hormone with potential to improve crop productivity. Indian J Plant Physiol 2006: 11; 1-12.

[3]   Vardhini BV, Rao SSR, Rao KVN. In: Ashwani Kumar S K, Sopory I K, Eds. Recent advances in plant biotechnology and its applications. New Delhi: International Publishing House Ltd 2008: pp. 133-39.

[4]   Xia X-J, Huang L-F, Zhau Y-H, Mao W-H, Shi K, Yu J-Q. Brassinosteroids promote photosynthesis and growth by enhancing activation of Rubisco and photosynthetic genes in *Cucumis sativus*. Planta 2009: 230; 1185-96.

[5]   Mitchell J W, Mandava NB, Worley JE, Plimmer JR, Smith MV. Brassins : a family of plant hormones from rape pollen. Nature 1970: 225; 1065-66.

[6]   Grove MD, Spencer GF, Rohwededer WK, *et al.* Brassinolide, a plant promoting steroid isolated from *Brassica napus* pollen. Nature 1979: 281; 121-24.

[7]   Bajguaz A. Isolation and characterization of brassinosteroids from algal cultures of *chlorella vulgaris* Beijernick (Trebouxiophyceae). J Plant Physiol 2009: 166; 1946-49.

[8]   Rao SSR, Vardhini BV, Sujatha E, Anuradha S. Brassinosteroids- A new class of phytohormones. Curr Sci 2002: 82; 1239-45.

[9]   Haubrick LL, Assmann SM. Brassinosteroids and plant function: some clues, more puzzles. Plant Cell Environ 2006: 29; 446-57.

[10]   Tao YZ, Zheng J, Xu ZM, Zhang XH, Zhang K, Wang GY. Functional analysis of ZmDWF1, a maize homolog of *Arabidopsis* brassinosteroids biosynthetic DWF1/DIM gene. Plant Sci 2004: 167; 743-51.

[11]   Nomura T, Nakayama N, Reid JB, Takeuchi Y, Yokota T. Blockage of brassinosteroid biosynthesis and sensitivity cause dwarfism in *Pisum sativum*. Plant Physiol 1997: 113; 31-7.

[12]   Strnad M, Kohout L. A simple brassinolide analogue 2α, 3α-dihydroxy-17β-(3-methyl butynyloxyl) 7-oxa-B-homo-5α androstan-6-one which induces bean second node splitting. Plant Growth Regul 2003: 40; 39-47.

[13]   Bishop GJ, Yokota T. Plant steroid hormones, brassinosteroids: Current highlights of molecular aspects on their synthesis, metabolism, transport, perception and response. Plant Cell Physiol 2001: 42; 114-20.

[14]   Zeng H, Tang Q, Hue X. *Arabidopsis* brassinosteroid mutants *del 2-1* and bin 2-1 display altered salt tolerance. J. Plant Growth Regul 2010: 29; 44-52.

[15]   Suzuki Y, Saso K, Fujioka S, *et al.* A dwarf mutant strain of *Pharbitis nil*, Uzukobito (kobito), has defective brassinosteroids biosynthesis. Plant J 2003: 36; 401-10.

[16]   Friedrichsen DM, Nemhauser J, Muramitsu T, *et al.* Three redundant brassinosteroids early response genes encode putative bHLH transcription factors required for normal growth. Genetics 2002: 162; 1445-56.

[17]   Asami T, Mink YK, Nagata N, *et al.* Characterization of brassinozole, a triazole-type brassinosteroid biosynthesis inhibitor. Plant Physiol 2000: 123; 93-100.

[18]   Nagata N, Asami T, Yoshida S. Brassinozole, an inhibitor of brassinosteroids biosynthesis, inhibits development of secondary xylem in cress plants (*Lepidium sativum*). Plant Cell Biol 2001: 42; 1006-11.

[19]   Shimida Y, Goda H, Nakamura A, Takatsuto S, Fujioka S, Yoshida S. Organ- specific expression of brassinosteroids-biosynthetic genes and distribution of endogenous brassinosteroids in *Arabidopsis*. Plant Physiol 2003: 131; 287-97.

[20]   Arteca JM, Arteca RN. Brassinosteroid-induced exaggerated growth in hydroponically grown *Arabidopsis* plants. Physiol Plant 2001: 112; 104-12.

[21]   Nakaya M, Tsukaya H, Murakami N, Kato M. Brassinosteroids control the proliferation of leaf cells of *Arabidopsis thaliana*. Plant Cell Physiol 2002: 43; 239-44.

[22]   Korableva NP, Platonova TA, Dogonadze MZ, Evsunina AS. Brassinolide effect on growth of apical meristems, ethylene production and abscisic acid content in potato tubers. Biol Plant 2002: 45; 39-43.

[23]   Ho MO. Brassinosteroids accelerate the rate of cell division in isolated protoplasts of *Petunia hybrida*. J. Plant Biotech 2003: 5; 63-7.

[24]   Lu Z, Huang M, Ge DP, *et al.* Effect of brassinolide on callus growth and regeneration in *Spartina patens* (Poaceae). Plant Cell Tiss Org Cult 2003: 73; 87-9.

[25]   Vardhini BV, Rao SSR. Effect of brassinosteroids on nodulation and nitrogenase activity in groundnut (*Arachis hypogaea*. L). Plant Growth Regul 1999: 28; 165-67.

[26]   Upreti KK, Murti GSR. Effects of brassinosteroids on growth, nodulation, phytohormones content and nitrogenase activity in French bean under water stress. Biol Plant 2004: 48; 407-11.

[27]   Hunter WJ. Influence of root applied epibrassinolide and carbenoxolone on the nodulation and growth of soybean (*Glycine max* L.) seedlings. J Agr Crop Sci 2001: 186; 217-21.

[28]   Vardhini BV, Rao SSR. Acceleration of ripening of tomato pericarp discs by brassinosteroids. Phytochem 2002: 61; 843-47.

[29]   Peng H, Tang XD, Feng HY. Effects of brassinolide on the physiological properties of litchi pericarp (*Litchi chinensis* cv nuomoci). Sci Hort 2004: 101; 407-16.

[30]   Ye Q, Zhu W, Li L, Zhang S, Yin Y, Ma H, Wang X. Brassinosteroids control male fertility by regulating the expression of key genes involved in *Arabidopsis* anther and pollen development. Proc Natl Acad Sci USA. 2010: 107; 6100-05.

[31]   Yang YH, Huang J, Ding J. Interaction between exogenous brassinolide, IAA and BAP in secondary metabolism of cultured *Onosma paniculatum* cells. Plant Growth Regul 2003: 39; 253-61.

[32]   Chang SC, Kim YS, Lee JY, *et al.* Brassinolide interacts with auxin and ethylene in root graviotropic response of maize (*Zea mays*). Physiol Plant 2004: 121; 666-73.

[33]   Tanaka K, Nakamura Y, Asami T, Yoshida S, Matsuo T, Okamato S. Physiological roles of brassinosteroids have a syngergistic relationship with giberellin as well as auxin in light –grown hypocotyls elongation. J Plant Growth Regul 2003: 22; 259-71.

[34]   Saka H, Fujii S, Imakawa AM, Kato N, Watanabe S, Nishizawa T, Yonekawa S. Effect of brassinolide applied at the meiosis and flowering stages on the levels of endogenous plant hormones during grain filling in rice plant (*Oryza sativa* L.). Plant Prod Sci 2003: 6; 36-42.

[35]   Arteca RN, Arteca JM. Effects of brassinosteroid, auxin and cytokinin on ethylene production in *Arabidopsis thaliana* plants. J Expt Bot 2008: 59; 3019-26.

[36]   Yu JQ, Huang LF, Hu WH, *et al.* A role for brassinosteroids in regulation in the photosynthesis in *Cucumis sativus*. J Expt Biol 2004: 55; 1135-43.

[37]   Deng Z, Zhang X, Tang W, *et al.* A proteomics study of brassinosteroid response in *Arabidopsis*. Mol Cell Proteomics 2007: 6; 2058-71.

[38]   Takematsu T, Takeuchi Y, Koguchi M. New plant growth regulators. Brassinolide and analogues, their biological effects and application in agriculture and biomass production. Chem Regul Plants 1983: 18; 1275-81.

[39]   Sasse JM. The case for brassinosteroids as endogenous plant hormones. In: Cutler HG, Yokota T, Adam G, Eds. Effect of brassinosteroids on protein synthesis and plant ultrastructure under stress conditions, Brassinosteroids – Chemistry, Bioactivity and Applications ACS Symp. Washington DC.: Am Chem Soc 1991: pp. 158-66.

[40]   Du L, Poovaiah BW. Ca^{2+}/calmodulin is critical for brassinosteroid biosynthesis and plant growth. Nature 2005: 437; 741-45.

[41]   Fujita M, Fujita Y, Noutoshi F, Narusaka Y, Yamaguchi-Shinozaki K, Shinozaki K. Cross talk between abiotic and biotic stress responses : a current view from the points of convergence in the stress signaling networks. Curr Opin Plant Biol 2006: 9; 436-42.

[42]   Sasse JM. Recent progress in brassinosterooid research. Physiol Plant 1997: 100; 697-701.

[43]   Krishna P. Brassinosteroid mediated stress responses. J Plant Growth Regul 2003: 22; 289-97.

[44]   Salcherk K, Bhalerao R, Koncz-Kalman Z, Koncz C. Control of cell elongation and stress responses by steroid hormones and carbon catabolic repression in plants. Philos Trans R Soc Lond B Biol Sci 1998: 1517-20.

[45]   Dhaubhadel S, Chaudhary S, Dobinson KF, Krishna P. Treatment of 24-epibrassinolide, a brassinosteroid, increases the basic thermotolerance of *Brassica napus* and tomato seedlings. Plant Mol Biol 1999: 40; 332-42.

[46]   Liu Y, Zhao Z, Si J, Di C, Han J, An L. Brassinosteroids alleviate chilling-induced oxidative damage by enhancing antioxidant defense system in suspension cultured cells of *Chorispora bungeana*. Plant Growth Regul 2009: 59; 207-14.

[47] Ozdemir F, Bor M, Demiral T, Turkan I. Effects of 24-epibrassinolide on seed germination, seedling growth, lipid peroxidation, proline content and antioxidative system of rice (*Oryza sativa* L) under salinity stress. Plant Growth Regul 2004: 42; 203-11.

[48] Dhaubhadel S, Browning KS, Gallie DR, Krishna P. Brassinosteroid functions to protect the translational machinery and heat-shock protein synthesis following thermal stress. Plant J 2002: 29; 681-91.

[49] Schnabl H, Roth U, Friebe A. Brassinosteroid induced stress tolerance in plants. Rec Res Develop Phytochem 2001: 5; 169-83.

[50] Janeczko A, Koscielniak J, Pilipowicz M, Szarek-Lukaszewsa G, Skoczowspi A. Protection of winter rape photosystem – 2 by 24-epibrassinolide under cadmium stress. Photosynthetica 2005: 43; 293-98.

[51] Ali B, Hasan SA, Hayat S, *et al.* A role for brassinosteroids in the amelioration of aluminium stress through antioxidant system in mung bean (*Vigna radiata* L. Wilczek). Environ Exp Bot 2008: 62; 153-59.

[52] Xia X-J, Zhang Y, Wu J-X, *et al.* Brassinosteroids promote metabolism of pesticides in cucumber. J Agric Food Chem 2009: 57; 8406-13.

[53] Xia X-J, Wang Y-J, Zhau Y-H, *et al.* Reactive oxygen species are involved in brassinosteroid-induced stress tolerance in cucumber. Plant Physiol 2009: 150; 801-14.

[54] Shen YY, Dai JY, Hu AC, Gu WL, He RY, Zheng B. Studies on physiological effects of brassinolide on drought resistance in maize. J Shenyang Agric Univ 1990: 21; 191-95.

[55] Li L, Staden JV. Effect of plant growth regulators on the drought resistance of two maize cultivars. South African J Bot 1998: 64,116-20.

[56] Li L, Staden JV, Jager AK. Effects of plant growth regulators on the anti- oxidant system in seedlings of two maize cultivars subjected to water stress. Plant Growth Regul 1998: 25; 81-7.

[57] Gonzalez-Suarez S, Gainza–Lezcano E. Physiological effects of synthetic brassinosteroid, DAA-6 on the invitro development of sugar plantlets. Revta Biol Habana 1997: 11; 53-60.

[58] Kagale S, Divi UK, Krochko JE, Keller WA, Krishna P. Brassinosteroids confers tolerance in *Arabidopsis thaliana* and *Brassica napus* to a range of abiotic stresses. Planta 2007: 225; 353-64.

[59] Divi UK, Rahaman T, Krishna P. Brassinosteroid –mediated stress tolerance in *Arabidopsis* shows interaction with abscisic acid, ethylene and salicylic acid. Plant Biol 2010: 10; 151-64.

[60] Jager CJ, Symons GM, Ross JJ, Reid JB. Do brassinosteroids mediate the water stress response? Physiol Plant 2008: 133; 417-25.

[61] Sairam RK, Shukla DS, Deshmuk PS. Effect of homobrassinolide seed treatment on germination, α- amylase activity and yield of wheat under moisture stress condition. Indian J Plant Physiol 1996: 1,141-44.

[62] Li L, Staden JV. Effects of plant growth regulators on the antioxidative system in seedlings of two maize cultivars subjected to water stress. Plant Growth Regul 1998: 25; 81-7.

[63] Rajashekaran LR, Blake TJ. New plant growth regulators protect photosynthesis and enhance growth under drought of jackpine seedling. J Plant Growth Regul 1999: 18; 175-81.

[64] Vardhini BV, Rao SSR. Amelioration of osmotic stress by brassinosteroids on seed germination and seedling growth of three varieties of sorghum. Plant Growth Regul 2003: 41; 21-31.

[65] Vardhini BV, Rao SSR. Influence of brassinosteroids on seed germination and seedling growth of sorghum under water stress. Indian J Plant Physiol 2005: 10; 381-5.

[66] Li KR, Wang HH, Han G, Wang Q, Fan J. Effects of brassinolide on the survival, growth and drought resistance of *Robinia pseudoacacia* seedlings under water-stress. New Forests 2008: 35; 255-66.

[67] Yuan GF, Jia CG, Li Z, *et al.* Effect of brassinosteroids on drought resistance and abscisic acid concentration in tomato under water stress. Sci Hort 2010: 126; 103-8.

[68] Schilling G, Schiller C, Otto S. Influence of brassinosteroids on organ relation and enzyme activities of sugar beet plants. In: Cutler HG, Yokota T, Adam G, Eds. Effect of brassinosteroids on protein synthesis and plant ultrastructure under stress conditions, Brassinosteroids – Chemistry, Bioactivity and Applications ACS Symp. Washington DC. Am Chem Soc 1991: pp. 208-19.

[69] Sairam RK. Effect of homobrassinolide application on plant metabolism and grain yield under irrigated and moisture -stress conditions of two wheat varieties. Plant Growth Regul 1994: 14; 173-81.

[70] Singh J, Nakamura S, Ota Y. Effect of epi-brassinolide on gram *(Cicer arietinum)* plants grown under water stress in juvenile stage. Indian J Agric Sci 1993: 63; 395-7.

[71] Kumawat BL, Sharma DD, Jat SC. Effect of brassinosteroid on yield and yield attributing characters under water deficit stress conditions in mustard (*Brassica juncea* L. Czern and Coss). Ann Bot Ludhiana 1997: 13; 91-3.

[72] Hnilicka F, Hnilickova H, Martinkova J, Blaha L. The influence of drought and the application of 24-epibrassinoide on the formation of dry matter and yield in wheat. Cer Res Commun 2007: 35; 457-60.

[73] Saygideger S, Deniz F. Effect of 24- epibrassinolide on biomass, growth and free proline concentration in *Spirulina platensis* (Cyanophyceae) under NaCl stress. Plant Growth Regul 2008: 56; 219-23.

[74] Prusakova LD, Chizhova SI, Ageeva LF, Golantseva EN, Yakovlev AF. Effects of epibrassinolide and ekost on the drought resistance and productivity of spring wheat. Agrokhimiya 2000: 3; 50-4.

[75] Fariduddin Q, Khanam S, Hasan SA, Ali B, Hayat S, Ahmad A. Effect of 28-homobrassinolide on the drought stress-induced changes in photosynthesis and antioxidant system of *Brassica juncea* L. Acta Physiol Plant 2009: 31; 889-97.

[76] Hnilicka F, Koudela M, Martinkova J, Hniclickova H, Hejnak V. Effect of water deficit and application of 24-epibrasasinolide on gas exchange in cauliflower plants. Sci Agricult Bohemica 2010: 41; 15-20.

[77] Zhang M, Zhai Z, Tian X, Duan L, Li Z. Brassinolide alleviated the adverse effect of water deficits on photosynthesis and the antioxidants of soybean (*Glycine max* L.). Plant Growth Regul 2008: 56; 257-67.

[78] Pustovaikova TN, Zhdonova NE, Zholkevich VN. Epibrassinolide increases plant drought resistance. Doklady Biochem Biophys 2001: 376; 36-8.

[79] Behnamnia M, Kalantari KH, Ziaie J. The effects of brassinosteroid on the induction of biochemical changes in *Lycopersicon esculentum* under drought stress. Turkish J Bot 2009: 33; 417-28.

[80] Xu HL, Shida A, Futatsuya F, Kumura A. Effects of epibrassinolide and abscisic acid on sorghum plants growing under water deficits. I. effects on growth and survival. Japan J Crop Sci 1994: 63; 671-5.

[81] Xu HL, Shida A, Futatsuya F, Kumura A. Effects of epibrassinolide and abscisic acid on sorghum plants growing under water deficits. II. Physiological basis for drought resistance induced by exogenous epibrassinolide and abscisic acid. Japan J Crop Sci 1994: 63; 676-81.

[82] Farooq M, Wahid A, Basra SMA, Islam-ud-Din. Improving water relations and gas exchange with brassinosteroids in rice under drought stress. J Agro Crop Sci 2009: 195; 262-9.

[83] Hathout TA. Salinity stress and its counteraction by the growth regulator (Brassinolide) in wheat plants (*Triticum aestivum* L- Cultivar Giza 157). Eygpt J Physiol Sci 1996: 20; 127-52.

[84] Vardhini BV, Rao SSR. Effect of brassinosteroids on salinity induced growth inhibition of ground nut seedlings. Indian J Plant Physiol 1997: 2; 156-7.

[85] Takeuchi Y. Studies on physiology and applications of brassinosteroids. Shokubutsu no Kogaku Chosetsu 1992: 27; 1-10.

[86] Ali B, Hayat S, Fariduddin Q, Ahmad A. 24-Epibrassinolide protects against the stress generated by salinity and nickel in *Brassica junceae*. Chemosphere 2008: 72; 1387-92.

[87] Zeng H, Tang Q, Hue X. *Arabidopsis* brassinosteroid mutants *del 2-1* and bin 2-1 display altered salt tolerance. J Plant Growth Regul 2010: 29; 44-52.

[88] Abd El- Fattah RI. Osmolytes –antioxidants behavior in *Phaseolus vulgaris* and *Hordeum vulgare* with brassinosteroid under water stress. American-Eurasian J Agric Environ Sci 2007: 2; 639-47.

[89] Hayat S, Ali B, Ahmad A. Response of *Brassica juncea*, to 28-homobrassinolide, grown from the seeds exposed to salt stress, J Plant Biol 2006: 33; 169-74.

[90] Amzallag GN, Gouloubinoff P. An Hsp90 inhibitor, geldanamycin, as a brassinosteroid antagonist: evidence from salt-exposed roots of *Vigna 24ynthes*. Plant Biol 2003: 5; 143-50.

[91] Amzallag GN. Brassinosteroid: a modulator of the developmental window for salt –adaption in *Sorghum bicolor*? Israel J Plant Sci 2004: 52; 1-8.

[92] Qayyam Z, Shahbaz M, Akram NA. Interactive effect of foliar application of 24-epibrassinolide and root zone salinity on morpho physiological attributes of wheat (*Triticum aestivum* L.). Int J Agric Biol 2007: 9; 584-9.

[93] Sasse JM, Smith R, Hudson I. Effect of 24-epibrassinolide on germination of seeds of *Eucalyptus camaldulensis* in saline conditions. Proc Plant Growth Regul Soc USA 1995: 22; 136-41.

[94] Anuradha S, Rao SSR. Effect of brassinosteroids on salinity stress induced inhibition of germination and seedling growth of rice (*Oryza sativa*.L). Plant Growth Regul 2001: 33; 151-3.

[95] Anuradha S, Rao SSR. Application of brassinosteroids to rice seeds (*Oryza sativa*.L) reduced the impact of salt stress on growth and improved photosynthetic levels and nitrate reductase activity. Plant Growth Regul 2003: 40; 29-32.

[96] Shakirova FM, Bezrukova MV. Effect of 24-epibrassinolide and salinity on the levels of ABA and lectin. Russ J Plant Physiol 1998: 45; 388-91.

[97] Anuradha S, Rao SSR. Effect of brassinosteroids on growth and antioxidant enzyme activities in rice seedlings under salt stress. Proc A P Acad Sci 2007: 11; 198-203.

[98] Arora N, Bharadwaj R, Sharma P, Arora HK. 28-Homobrassinolide alleviates oxidative stress in salt treated maize (*Zea mays* L) plants. Brazil J Plant Physiol 2008: 20; 153-7.

[99]   Kilic S, Cavusoglu K, Kaber K. Effects of 24-epibrassinolide on salinity stress induced inhibition of seed germination, seedling growth and leaf anatomy of barley. SDU Fen Edebiyast Fakult Fen Dergisi 2007: 2; 41-51.

[100]   Hamada K. Brassinolide in crop cultivation. In: McGregor, P., Ed. Plant Growth Regulators in Agriculture. Taiwan: FFTC 1986: Vol. 1; pp 190-6.

[101]   Kulaeva ON, Burkhanova EA, Fedina AB, *et al.* Brassinosteroids in regulation of protein synthesis in wheat leaves. Dokl AN USSR 1989: 305; 1277-79.

[102]   Bokebayeva GA, Khripach VA. Effect of 24-epibrassinolide on seed germination and growth of barley plants under salt stress. In: Brassinosteroids-Biorational, ecologically safe regulators of growth and productivity of plants. 3rd ed. Minsk: Byelorussian Sci 1993: p.21.

[103]   Shahbaz M, Ashraf M. Influence of exogenous application of brassinosteroids on growth and mineral nutrients of wheat (*Triticum aestivum* L) under saline conditions. Pak J Bot 2007: 39; 513-22.

[104]   Shahbaz M, Ashraf M, Athar HR. Does exogenous application of 24-epibrassinolide ameliorate salt induced growth inhibition in wheat (*Triticum aestivum* L.)? Plant Growth Regul 2008: 55; 51-64.

[105]   Houmili SM, Denden ME, El Hadj SB. Induction of salt tolerance in pepper (*Capsicum annum*) by 24-epibrassinolide. Eur Asia J Bio Sci 2008: 2; 83-90.

[106]   Hayat S, Ali B, Ahmad A. Response of *Brassica junceae* to 28-homobrassinolide grown from seeds exposed to salt stress. J Plant Biol 2006: 33; 169-74.

[107]   Kulaeva ON, Burkhanova EA, Fedina AB, *et al.* In: Cutler HG, Yokota T, Adam G, Eds. Effect of brassinosteroids on protein synthesis and plant ultrastructure under stress conditions, Brassinosteroids – Chemistry, Bioactivity and Applications ACS Symp. Washington DC. Am Chem Soc 1991: pp. 141-55.

[108]   Ali B, Hayat S, Ahmad A. 28-Homobrassinolide ameliorates the saline stress in *Cicer arietinum* L. Environ Exp Bot 2007: 59; 217-23.

[109]   Hayat S, Ali B, Hasan SA, Ahmad A. Effect of 28-homobrassinolide on salinity-induced changes in *Brassica juncea*. Turkish J Biol 2007: 31; 141-6.

[110]   Ali Q, Athar HR, Ashraf M. Modulation of growth, photosynthetic capacity and water relations in salt stressed wheat plants by exogenously applied 24-epibrassinolide. Plant Growth Regul 2008: 56; 107-16.

[111]   Nunez M, Mazzafera P, Mazzora LM, Sigueira WJ, Zullo M. Influence of a brassinosteroid analogue on antioxidant enzymes in rice grown in culture medium with NaCl. Biol Plant 2003: 47; 67-70.

[112]   Ali Q, Abd El-Fattah RI. Osmolytes –antioxidants behavior in *Phaseolus vulgaris* and *Hordeum vulgare* with brassinosteroid under water stress. J Agro 2006: 5; 167-74.

[113]   Khrustaleva LI, Pogorilaya EV, Golovnina Yu-M, Andreeva GN. Effect of epibrassinolide on mitotic activity and the frequency of chrosome aberrations in barley root tip cells under salt stress. Sel'skokhozyaistvennaya-Biol 1995: 5; 69-73.

[114]   Tabur S, Demick K. Cytogenetic response of 24-epibrassinolide on the root meristem cells of barley seeds under salinity. Plant Growth Regul 2009: 58; 119-23.

[115]   Hayat S, Ali B, Hasan SA, Ahmad A. Effect of 28-homobrassinolide on salinity induced changes in growth, ethylene and seed yield in mustard. Indian J Plant Physiol 2007: 12; 207-11.

[116]   Abraham E, Rigo G, Szekely G, Nagy R, Koncz C, Szabados L. Light-dependent induction of proline biosynthesis by abscisic acid and salt stress is inhibited by brassinosteroid in *Arabidopsis*. Plant Mol Biol 2003: 51; 363-82.

[117]   Avalbaev AM, Yuldashov RA, Fatkhutdinova RV, Urusov FA, Safufdinova, YV, Shakirova FM. The influence of 24-epibrassinolide on the hormonal status of wheat plants under sodium chloride. Appl Biochem Microbiol 2010: 46; 99-102.

[118]   Yokota T, Takahashi N. In: Bopp M, Ed. Chemistry, physiology and agricultural applications of brassinolide and related steroids as plant growth substances. Berlin/Heildberg: Springer-Verlag 1986: pp. 129-38.

[119]   Bajguz A, Hayat S. Effects of brassinosteroids on plants exposed or subjected to various stresses. Plant Physiol Biochem 2009: 47; 1-8.

[120]   Mandava NB. Plant growth promoting brassinosteroids. Ann Rev Plant Physiol Plant Mol Biol 1988: 39; 23-52.

[121]   Sczekers M, Koncz C. Biochemical and genetic analysis of brassinosteroid metabolism and function in *Arabidopsis*. Special Issue: *Arabidopsis thaliana*. Plant Physiol Biochem 1998: 36; 145-55

[122]   Kamuro Y, Takatsuto S. Practical application of brassinosteroids in agricultural fields. In: Sakurai A, Yokota T, Clouse SD, Eds. Brassinosteroids-Steroidal plant hormones. Tokyo: Springer-Verlag 1999: pp. 223-41.

CHAPTER 3

Brassinosteroid-Driven Stimulation of Shoot Formation and Elongation: Application in Micropropagation

A. B. Pereira-Netto[1*], L. R. Galagovsky[2] and J. A. Ramirez[2]

[1]*Department of Botany-SCB, Centro Politécnico-Parana Federal University, C.P. 19031 Curitiba, PR, 81531-970, Brazil and* [2]*Department of Organic Chemistry and UMYMFOR (CONICET-FCEyN), School of Exact and Natural Sciences, University of Buenos Aires, Argentina*

Abstract: Brassinosteroids (BRs) comprise a specific class of low-abundance, natural polyhydroxy steroidal lactones and ketones now recognized as a new class of phytohormones. These steroids of ubiquitous occurrence in plants are known to stimulate stem elongation and to control apical dominance as well. In this chapter, we describe the use of BRs to significantly improve protocols for micropropagation of woody species, more specifically, the marubakaido apple rootstock and a clone of *Eucalyptus*, through the stimulation of shoot elongation and formation of new shoots. It is also shown in this chapter that these BR-induced changes in shoot architecture do not result from changes in the endogenous levels of any single metabolite and do not rely on broad changes in the metabolite profile, as well.

Keywords: 22-hydroxylated BRs, 5F-28-homocastasterone, brz220, *E. grandis* X *E. urophylla* hybrid, marubakaido, metabolomics, multiplication rate.

INTRODUCTION

Plant growth is accomplished by orderly cell division and elongation. Brassinosteroids (BRs), the most recently discovered class of plant growth substances, induce a broad spectrum of responses, however, growth stimulation through cell elongation and cell division is a major biological effect of BRs. A broad spectrum of responses has been shown to be elicited by exogenous application of BRs to either intact plants or excised organs. Exogenous application of BRs at nanomolar to micromolar concentrations to plants stimulates cell elongation and division, and vascular differentiation [1-3], all-important to allow shoot elongation.

Shoot proliferation is a powerful tool to improve micropropagation techniques, especially for woody species, in which new branch formation and elongation is typically a constrain for efficient micropropagation. In this chapter, we describe our experience on the use of 28-homocastasterone (1) and two synthetic 5α substituted analogs: 5α-fluorohomocastasterone (2) and 5α-hydroxyhomocastasterone (3) (Fig. 1) [(22R,23R)-5α-fluoro-2α,3α,22,23-tetrahydroxystigmastan-6-one and (22R,23R)-2α,3α,5α,22,23-pentahydroxystigmastan-6-one, respectively] for the stimulation of shoot proliferation in micropropagation systems for two tree species, the marubakaido apple rootstock and a hybrid between *Eucalyptus grandis* and *E. urophylla*.

Effect of 28-HCS, 5F-HCS and 5OH-HCS on the *In Vitro* Multiplication Rate and Shoot Formation on the Marubakaido Apple Rootstock

Marubakaido (*Malus prunifolia*, Willd, Borkh) is one of the most widely used apple rootstock in several countries. Similarly to what often occurs with other woody species, *in vitro* multiplication rates (MRs) reported for the marubakaido apple rootstock (*Malus prunifolia* (Willd.) Borkh cv. Marubakaido) are low, typically at the 4-5 new shoots per explant range [4]. This low MR makes the micropropagation techniques available for this rootstock barely feasible for commercial purposes.

*Address correspondence to A. B. Pereira-Netto: Department of Botany-SCB, Centro Politécnico-Parana Federal University, C.P. 19031 Curitiba, PR, 81531-970, Brazil; E-mail: apereira@ufpr.br

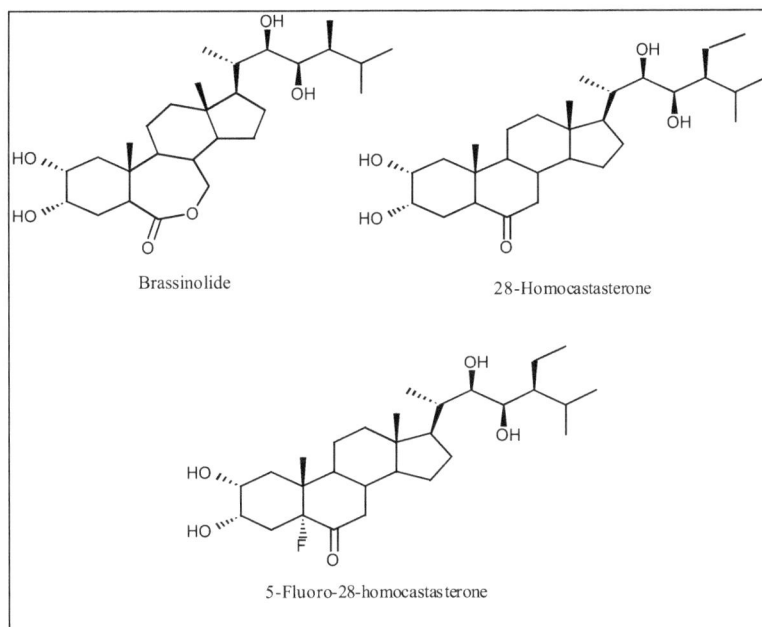

Figure 1: Structural formulae of brassinolide, 28-Homocastasterone and 5-Fluoro-28-homocastasterone.

28-homocastasterone (28-HCS) is a naturally occurring BR in plants widely employed in field trials because of its greater synthetic accessibility compared to typically more active BRs, such as brassinolide (BL) [1,5-6]. The rapid metabolism of natural BRs in plants is a major limitation for the broader use of BRs in agriculture. Thus, the synthesis of new analogs, besides enabling studies of structure-activity relationships, biosynthesis and metabolism of BRs, is a possible way of overcoming the rapid metabolism of natural BRs in plants once new analogs are demonstrated to be more difficult to be metabolized by plants. In order to enlarge studies on the effects of BRs analogues on bioactivity, Ramirez and co-workers [7] synthesized various new BRs analogues, some of which were fluoro derivatives of 28-HCS. Paralleling tests carried out in Argentina, in which antiviral (Chapter **5**, in this book), antiherpetic and anti-inflamatory activity (Chapter 6, in this book) of new HCS and other BR derivatives were tested, the effect of new HCS derivatives were tested in micropropagation systems for woody species, in Brazil.

Increase on the multiplication rate (MR, number of neoformed branches ≥ 15 mm in length, the minimum length suitable for propagation purposes) for *in vitro*-grown *Malus prunifolia* shoots was associated with leaf application of 5α-fluorohomocastasterone (5F-HCS) (compound **2**, Fig. **1**) in the 500 to 1000 ng per shoot range [8]. Treatment of shoots with 500 ng per shoot of 5F-HCS resulted in MR of 7.3, a 112% increase on MR, compared to untreated shoots. Thus, this 5F-HCS-induced enhancement in the multiplication rate consists in an effective way to improve the micropropagation technique for the apple rootstock. Virtually no change on multiplication rates was found for shoots treated with either 28-homocastasterone (compound **1**, Fig. **1**) (28-HCS) or 5α-hydroxyhomocastasterone (compound **3**, Fig. **1**) (5OH-HCS).

A significant 15% increase in the number of main shoots (branches originated directly from the initial explant formed during the culture cycle) was found for shoots treated with 500 ng of 5F-HCS. Since mutants defective on the biosynthesis or on the signal transduction pathway of BRs typically display reduced apical dominance, it was somewhat surprisingly to find that shoots treated with 500 ng of 5F-HCS also presented a significant (p=0.05) 238% and 250 % increase, respectively, for the number of primary lateral shoots (branches originated from the main branches) and for the number of secondary lateral branches (branches originated from the primary lateral branches), measuring at least 15 mm in length (suitable for micropropagation). Thus, the increase on the multiplication rate found for shoots treated with 500 ng per shoot of 5F-HCS was mainly due to an increase in both, number of primary and secondary lateral branches.

Differently from 5F-HCS, which induced remarkable changes in shooting pattern of *in vitro*-grown marubakaido shoots, 28-HCS and 5OH-HCS applications resulted in no significant change in any of the features evaluated. Summarizing, compounds 1, 2 and 3 presented contrasting effects on the architecture of *in vitro*-grown marubakaido shoots. Compound 2 stimulated shoot proliferation, through shoot elongation, but especially through an increase on lateral shooting, which resulted in significantly enhanced multiplication rate for the marubakaido apple rootstock. Since effects of plant hormones and analogues on plant growth and metabolism depend on the extents to which these molecules satisfy the structural requirements of the receptors and enzymes, the differential effects we describe for 28-HCS and two of their analogs provide clues to probe into the signal transduction pathways and metabolism of BRs. Once the substitution of the hydrogen atom by a fluorine atom in a carbon-hydrogen bond is known to cause significant increase in electronegativity and hydrogen bonding potential [9], data described for the marubakaido rootstock suggests that increased electronegativity and/or hydrogen bonding potential at carbon 5 of 28-HCS may improve the ability of 28-HCS to satisfy the structural requirements of the BR receptors, triggering more efficiently the BR signal transduction pathway, leading to branching in our system.

EFFECT OF 5F-HCS ON THE SHOOT ELONGATION OF MARUBAKAIDO

Progressive enhancement of elongation of main and primary lateral shoots of *in vitro*-grown marubakaido shoots was associated with increased doses of 5F-HCS until 0.5 μg per shoot. The use of 5F-HCS at doses over 0.5 μg per shoot led to reduced shoot length, compared to the peak stimulation of shoot elongation found at 0.5 μg per shoot. Continuous reduction on elongation of main and primary lateral branches was found for shoots treated with increased doses of 5F-HCTS over 0.5 μg per shoot. Treatment of marubakaido shoots with 0.5 μg per shoot of 5F-HCS resulted in significant (p=0.05) 21 and 30% enhancement of shoot elongation for main and primary lateral shoots, respectively, demonstrating that primary lateral shoots were more responsive to the BR treatment, towards shoot elongation, compared to the main shoots. For the secondary lateral shoots, no significant change in elongation was found for BR-treated shoots at any of the doses tested.

Analysis of the Possible Involvement of Ethylene in the 5F-HCS-Induced Change in the Pattern of Shoot Elongation of Marubakaido

BRs have been demonstrated to stimulate 1-amino-cyclopropane-1-carboxylic acid (ACC), the immediate precursor of ethylene in higher plants, and ethylene biosynthesis in systems such as the primary roots of maize (*Zea mays*), in which exogenously applied BL enhances ethylene release and ACC oxidase activity, in a dose-dependent manner [10]. Furthermore, in etiolated mung bean hypocotyl segments, stimulation of ethylene biosynthesis, due to increased ACC synthase activity, has been found after BL treatment [11]. Indeed, in the past years, molecular genetics studies have been able to identify specific genes of the ethylene biosynthetic pathway activated by BRs, helping to understand how BRs affect ethylene biosynthesis at the gene expression level. For example, experiments carried out with etiolated plantlets of *cin5*, a loss-of-function *Arabidopsis* mutant defective in ACS5, a member of the ACC synthase gene family, have suggested that BR-induced stimulation of ethylene biosynthesis is at least in part dependent on accumulation of the ACS5 isoform [12].

Time course analysis of ethylene release by the marubakaido rootstock demonstrated that a progressive increase in rate of ethylene release occurred between Day 1 (the beginning of the culture cycle) and Day 8, for shoots treated with 5F-HCS, and between Day 1 and Day 4 for shoots not treated with 5F-HCS [13]. For either Day 4 or Day 8 shoots treated with 5F-HCS typically presented higher rates of ethylene release when compared to untreated shoots. After ethylene release peaked, a decrease in the rate of release was observed until it stabilized around four nmol g^{-1} day^{-1} for shoots either treated or not with 5F-HCS.

Increasing concentrations of 1-amino-cyclopropane-1-carboxylic acid (ACC), the immediate precursor of ethylene in higher plants, in the culture medium resulted in decreased length of main, primary lateral and secondary lateral shoots [13]. These decreases in length become significant (P=0.05) at 3.1525, 50 and 200 μM ACC, respectively, for secondary lateral, main and primary lateral shoots, when compared to the control (0 μM ACC).

Similarly to what was found for shoots grown in ACC-containing media, progressive decrease of elongation was found for main and primary lateral shoots of shoots grown in atmosphere enriched with increased concentrations of ethylene at the 10 to 60 μmol l^{-1} range. These decreases were statistically significant (P=0.05) for all of the ethylene concentrations tested. However, for secondary lateral shoots significant inhibition of elongation was only found for shoots grown at 40 and 60 μmol l^{-1} ethylene, indicating that secondary lateral shoots are less sensitive to low ethylene concentrations, toward inhibition of elongation, when compared to main and primary lateral shoots. However, differently from what was found for main and primary lateral shoots, complete inhibition of secondary lateral shoot elongation was found for shoots grown in atmosphere enriched with 40 and 60 μl l^{-1} ethylene.

Ethylene is well known for a long time to inhibit organ elongation, such as in hypocotyls of etiolated seedlings, in numerous plant species [14, 15]. However, stimulation of elongation has been found for light-grown seedlings, as well. In *Arabidopsis*, for example, seedlings grown in the light in either an atmosphere enriched with 10 ppm ethylene or in the presence of 50 μM ACC in a low nutrient culture medium presented longer hypocotyls, compared to its counterparts grown in ethylene-free atmosphere or ACC-free culture medium [16]. Furthermore, ethylene is known for quite some time to stimulate elongation of water logged plants [17]. Concluding, the BR-stimulated marubakaido shoot elongation is paralleled by an enhancement of ethylene release from the shoots. However, enrichment of either the culture medium with ACC or the internal vessel atmosphere with ethylene resulted in inhibition of shoot elongation, which provides evidence that the stimulation of shoot elongation observed for 5F-HCS-treated shoots in this study is not, at least directly, related to the BR-induced enhancement in ethylene release rate.

The results presented in this chapter show a new application for BRs in horticulture. The C-5 fluoro derivative of 28-HCS-induced shoot proliferation is an effective method to enhance the *in vitro* multiplication rate for marubakaido, which significantly contributes to make commercially feasible the micropropagation technique for this apple rootstock, besides increasing the availability of certified, virus-free propagules. In addition, the 5F-HCS-driven enhancement of shooting described in this chapter is potentially useful to improve micropropagation techniques for other plant species as well, especially woody species, in which shoot elongation is typically a constrain for efficient micropropagation.

BL-Induced Changes in Shoot Morphology of Marubakaido

BL, the most potent naturally occurring BR [18], differentially affected formation and further elongation of main and primary lateral shoots of *in vitro*-grown shoots of the marubakaido apple rootstock. Analysis of the effect of BL on shooting demonstrated that BL more effectively stimulated formation and further elongation of primary lateral shoots compared to main shoots, which resulted in reduced apical dominance. Progressive increase in shoot length was related to increased doses of BL. A statistically significant increase of 12% and 25%, respectively, for the main and primary lateral shoot length was found for shoots treated with 1.25 μg.shoot^{-1} BL, compared to untreated shoots. BL was also shown to more effectively stimulates shooting in *M. prunifolia*, compared to 28-HCS [19 and unpublished results]. However, since BL is thought to be the most biochemically active natural BR [20], it is somewhat surprising to find that BL was less effective at inducing shooting, compared to 5F-HCS [8, 13]. Nevertheless, BL was more effective, compared to 5F-HCS at stimulating main shoot formation. The reason(s) why BL is less effective at stimulating shoot elongation and formation of primary lateral shoots, compared to 5F-HCS, is(are) complex. It is possible that the differential effects of BRs and their analogues on shoot formation and elongation might depend on the extent to which these molecules satisfy the structural requirements of BR receptors. However, other possibilities such as an eventually higher susceptibility of BL to inactivation by BAS1, a BR inactivating gene [21-23], compared to 5F-HCS, can not be ruled out.

Time Course Analysis of the Metabolite Profile of 5F-HCS-Treated and Untreated Marubakaido Shoots

Although the molecular mode of action of BRs has been largely studied at the gene expression level [24], little is known about the downstream signaling elements, which makes it difficult to establish a connection between BR sensing and actual physiological changes. In *M. prunifolia*, shoots treated with 5F-HCS were shown to present increased shoot formation [8] and elongation [13]. Metabolome analysis was used to help to unravel how 5F-HCS stimulates formation and elongation of shoots in the marubakaido rootstock. The metabolite profiles of marubakaido shoots were analysed before the start of the treatment with 5F-HCS and one and four weeks after the treatment. A total of 120 different metabolites were detected, of which 102 were structurally identified, in marubakaido shoots and 5F-HCS-treated marubakaido shoots. Analysis of the metabolite profile of 5F-HCS-treated and untreated marubakaido shoots showed that the number and abundance of compounds varied with culture time. Fructose and glucose were higher in 5F-HCS-treated shoots compared to untreated shoots by the end of the culture cycle (4 weeks after the treatment). No significant change in the identified amino acids or fatty acids was shown to be related to the 5F-HCS treatment. Among the organic acids, the only significant change was an increase of the putatively identified gulonic acid found for 5F-HCS-treated shoots, compared to untreated shoots. In fruits of the *Lycopersicum esculentum d^x* BR-deficient mutant, various sugars, including fructose and glucose were significantly reduced in fruits of the mutant, compared to the wildtype, at several stages of the fruit development [25]. The d^x mutant contains no detectable castasterone or BL (bioactive BRs) in vegetative tissues [26]. However, d^x fruits produce considerable amounts of bioactive BRs [27]. Because BR-application to leaves normalized levels of sugars such as fructose and glucose in the mutant, bioactive BR in leaves are considered to be required for sugar accumulation in tomato fruits [25]. When taken together, data for the d^x mutant and data from the metabolomic analysis of 5F-HCS-treated marubakaido shoots suggest that glucose and fructose accumulation is positively related to BR availability in shoots of *M. prunifolia* and source tissues [28] in *L. esculentum*. The metabolite profile of 5F-HCS-treated and untreated shoots indicate that none of the analysed metabolites can explain, by itself, the 5F-HCS-induced changes in shoot morphology of the *in vitro*-grown marubakaido shoots. Thus, the 5F-HCS-induced changes in shoot architecture, *i.e.* shoot elongation and formation of new shoots, in marubakaido shoots very likely involve a variety of different mechanisms and consequently do not result from changes in the endogenous levels of any single metabolite. In addition, the changes in the metabolite profile found for shoots treated with 5F-HCS, compared to untreated shoots, indicate that the 5F-HCS-induced growth stimulation, for example the 238% increase in the number of primary lateral shoots, does not rely on broad changes in the metabolite profile.

Effect of Brz220 on the Micropropagation of Marubakaido

Mutants deficient in BR biosynthesis have significantly contributed to increased knowledge about BRs and their actions. However, the use of specific biosynthesis inhibitors is an alternative way for the determination of physiological functions of BRs especially for trees, as no BR-deficient tree mutant has been identified thus far. BR-deficient mutants of several plants exhibit strong dwarfism with curly and dark-green leaves in the light [20], features typically attributed to a constitutive decrease of endogenous BRs (reviewed in [23]). Treatment of marubakaido shoots with Brz 220 [29], a potent and selective inhibitor of the hydroxylation of C22- of the side chains of BRs [30-34], demonstrated that Brz 220 affects differentially elongation of main and primary lateral shoots in marubakaido. Progressive inhibition of main shoot elongation was found for shoots treated with increasing doses of Brz 220. Surprisingly, a slight increase in primary lateral shoot elongation was found for Brz 220-treated shoots, except when Brz 220 was used at the highest dose (5µg.shoot^{-1}), where no change on shoot elongation was found. Treatment of *M. prunifolia* shoots with 0.2 µg.shoot^{-1} Brz 220 resulted in an increase of 16 and 67 % in the formation of main and primary lateral shoots, respectively, although these increases were not statistically significant. Regardless Brz 220 was more effective towards the enhancement of primary lateral shoot formation, compared to main shoots, the Brz 220-induced stimulation of shooting was not surprising as the BR-mutant bri1-1 is extremely shooted in older *Arabidopsis* plants [35].

Effect of Brz 220 on the Endogenous Levels of Sterols and Intermediates of the BL Biosynthetic Pathway in Marubakaido

The importance of BRs in a broad range of developmental processes implies the existence of mechanisms controlling both, endogenous levels and distribution of BRs in the target cells or tissues. An integrated understanding of the role of BRs in plant development requires a deep understanding of the way plants modulate the endogenous levels of, and responsiveness to, endogenous BRs [36].

GC-MS chemical analysis carried out on Brz 220-treated and untreated marubakaido shoots allowed the identification and quantification of teasterone (TE), typhasterol (TY), castasterone (CS), 6-deoxocathasterone (6-deoxoCT), 6-deoxoteasterone (6-deoxoTE), 3-dehydro-6-deoxoteasterone (6-deoxo3DT), 6-deoxotyphasterol (6-deoxoTY), and 6-deoxocastasterone (6-deoxoCS) as endogenous BRs in marubakaido. However, neither brassinolide (BL) nor cathasterone (CT) were identified. The treatment with increasing doses of Brz 220 led to a progressive decline in the endogenous levels of 6-deoxoTY, CS, and especially, 6-deoxoCT. The treatment with 10 µg of Brz 220 also led to a decline in the endogenous levels of other BRs such as 6-deoxoCS and TY. The Brz 220-related decline in the endogenous levels of 6-deoxoTY, CS, 6-deoxoCT, 6-deoxoCS and TY give support to previous reports demonstrating that Brz treatments diminish the levels of BRs downstream of the 22-hydroxylated BRs [32].

When taken together, data on the effects of BRs and Brz on both, shoot elongation and formation, along with data on the influence of Brz on the endogenous levels of BRs, indicate that shooting in marubakaido is under the control of changes in the endogenous pool of BRs.

The results presented in this chapter demonstrate that shooting management, *i.e.*, control in the allocation of growth among the various types of shoots, can be achieved at the biochemical level in marubakaido through induced changes in the endogenous levels of BRs. Additionally, these findings have potential practical applications, other than the improvement of a micropropagation technique for the marubakaido apple rootstock, in horticulture, for the control of excessive vegetative growth or conversely, to promote shooting in producing orchards; in forestry, for the inhibition of lateral growth; and in forestry and horticulture, for the improvement of micropropagation techniques for clonal propagation of the usually hard-to-propagate woody species.

Comparative Effects of 28-Homocastasterone and its 5α-Monofluoro Derivative on the Multiplication Rate of an *in vitro*-Grown Hybrid between *Eucalyptus Grandis* and *E. Urophylla*

Because of their fast growth and wide range of adaptability, eucalyptus are the most widely employed species for reforesting in tropical and subtropical regions of the world. Due to their superior growth and quality of pulp, the *Eucalyptus grandis* X *E. urophylla* hybrid is rapidly replacing other forest species and *Eucalyptus* hybrids in managed forest plantations in Africa and South/Central Americas.

Shoots of an *in vitro*-grown *Eucalyptus grandis* X *E. urophylla* hybrid dipped in acetone containing known amounts of 28-HCS displayed enhanced multiplication rate, enhanced formation and further elongation of main shoots (shoots originated directly from the initial explant), and reduced formation and elongation of primary lateral (shoots originated from the main shoots) shoots [37]. Multiplication rate raised significant (P=0.05) 34% for shoots treated with 10 mg.l⁻¹ 28-HCS, compared to shoots treated with acetone, only. Further increase in 28-HCS concentrations led to reduction in the multiplication rate. Treatment of shoots with increased 28-HCS concentrations (4 and 10 mg.l⁻¹) also led to an increase in the number of main shoots formed, being this effect statistically significant (P=0.05) only for the 10 mg.l⁻¹ concentration. 28-HCS concentrations beyond 10 mg.l⁻¹ did not lead to further increase in the number of main shoots formed. In contrary, the use of 28-HCS at 62.5 mg.l⁻¹ resulted in reduction in the number of main shoots formed, compared to the control (shoots treated with acetone, only).

Stimulation of main shoot elongation was found for shoots immersed in solutions of 28-HCS for all concentrations tested, compared to shoots treated with acetone (control) only. However, this stimulation was statistically significant (P=0.05) only for the 10 mg.l⁻¹ concentration. Conversely to what was found for main shoots, treatment with enhanced concentrations of 28-HCS led to inhibition of primary lateral shoots elongation, although the effect was statistically significant (P=0.05) only for the 25 mg.l⁻¹ concentration.

Monofluoro analogues of gibberellins are known for a long time to show higher biological active in assays such as the lettuce hypocotyls elongation, when used in concentrations within the 10-5 to 10-7M range, when compared to their parental counterparts [38]. Differently from other plant hormones such as gibberellins, reports on the effects of fluoro-BRs are limited. The finding that 28-HCS was able to stimulate formation and elongation of main shoots, and consequently to enhance *in vitro* multiplication rate of the *E. grandis* X *E.*

urophylla hybrid, led us to investigate if a 5 α-fluoro derivative of HCS (5F-HCS) might be able to amplify the stimulatory effect of 28-HCS on formation and elongation of main shoots. 5F-HCS induced responses remarkably different from those induced by the parental counterpart. For shoots treated with increased concentrations of 5F-HCS, it was observed a progressive decrease in the multiplication rate in consequence of reduced formation of both, main and primary lateral shoots. These reductions in MR were related to progressive reduction in the number of main shoots useful for micropropagation (measuring at least 15 mm in length) found for shoots treated with increased concentrations of 5F-HCS, although this reduction was not statistically significant (P=0.05) for any of the 5F-HCS concentrations used in the trial. For shoots treated with the fluoro derivative of 28-HCS, no significant (P=0.05) change in main shoot elongation was found for any of the 5F-HCS concentrations used in the trial, although a light trend towards inhibition of elongation is likely to be in place. Progressive enhancement in the average length of primary lateral shoots was found for shoots treated with increased concentrations of 5F-HCS up to 25 mg.l^{-1}. However, this promotive effect was statistically significant (P=0.05) only for the 25 mg.l^{-1} concentration. Further increase in 5F-HCS concentration (62.5 mg.l^{-1}) led to a decline in the average length of primary lateral shoots, compared to the peak stimulation (23%) of shoot elongation found at 25 mg.l^{-1} 5F-HCS. The extent in which 5F-HCS stimulated primary lateral shoots elongation did not differ significantly from the extent in which 28-HCS stimulated main shoots elongation.

Both, 28-HCS and its fluoro derivative negatively affected formation of primary lateral shoots. Increased concentrations of 28-HCS and 5F-HCS essentially led to progressive reduction in the number of primary lateral shoots formed during the culture cycle. However, at the 25 mg.l^{-1} concentrations, 28-HCS was significantly (P=0.05) more effective in inhibiting primary lateral shoots formation, compared to 5F-HCS. Thus, the increase on the muliplication rate found for shoots treated with 10 mg.l^{-1} 28-HCS was due to the 28-HCS-driven enhancement in the formation of new main shoots.

The reason(s) why 28-HCS and 5F-HCS induce antagonistic responses towards either elongation of main and especially primary lateral shoots, or formation of main shoots in the *Eucalyptus* hybrid does not appear to be straightforward. A possible formation of a hydrogen bond involving fluorine (please see further discussion below) and a consequent reduced ability to bind to the BR receptor might explain the inability of 5F-HCS to stimulate elongation and formation of main shoots in the *E. grandis* X *E. urophylla* hybrid. Although BRs are known to stimulate elongation of young tissues (reviewed in [39]), these plant growth regulators have also been reported to inhibit shoot elongation. For example, BRs such as BL and 24-epiBL have previously been shown to inhibit shoot elongation in species such as rice (*Oryza sativa*) [40] and pea (*Pisum sativum*) [41], respectively.

Considering that effects of BRs and analogues on shoot formation and elongation depend on the extents to which these molecules satisfy the structural requirements of the receptors and/or enzymes, the differential responses found for 28-HCS and its 5α-fluoro substituent-treated shoots of *Eucalyptus* described in this chapter suggests either different metabolic routes, different chemical stability or different receptor sites for 28-HCS and 5F-HCS in the *E. grandis* X *E. urophylla* hybrid, and also in the marubakaido rootstock.

The enhancement in the multiplication rate found for 28-HCS-treated shoots described in this chapter demonstrates that BRs can be used for the improvement of protocols for *Eucalyptus* micropropagation (Patent BR PI0403642-5). In addition, the differential effect of BRs on the morphogenetic potential of main and primary lateral shoots and, consequently apical dominance, indicates that BRs might be useful to manage shooting in field-grown trees, especially in the genera *Malus* and *Eucalyptus*.

ACKNOWLEDGEMENTS

The Brazilian author is thankful to CNPQ-Brazil for financial support. Argentine authors are greatful to the University of Buenos Aires and UBA-CONICET for financial support.

REFERENCES

[1]   Mandava N B. Plant growth-promoting brassinosteroids. Ann Rev Plant Physiol Plant Mol Biol 1988; 39: 23-52.
[2]   Clouse S D, Sasse J M. Brassinosteroids: essential regulators of plant growth and development. Ann Rev Plant Physiol Plant Mol Biol 1998; 49: 427-51.

[3]   Sasse J. Physiological actions of brassinosteroids. In: Sakurai A, Yokota T, Clouse SD, Eds. Brassinostreoids: Steroidal Plant Hormones. Tokyo: Springer-Verlag 1999; pp. 137-61.

[4]   Nunes JCO, Barpp A, Silva FC, Pedrotti EL. Micropropagation of rootstocks "marubakaido" (*Malus prunifolia*) through meristem culture. Rev Bras Frutic 1999; 21: 191-95.

[5]   Fujioka S, Sakurai A. Biosynthesis and metabolism of brassinosteroids. Physiol Plant 1997; 100: 710-71.

[6]   Baron DL, Luo W, Janzen L, Pharis RP, Back TG. Structure-activity studies of brassinolide B-ring analogues. Phytochem 1998; 49: 1849-58.

[7]   Ramírez JA, Gros E, Galagovsky L. Effect on bioactivity due to C-5 heteroatom substituents on synthetic 28-Homobrassinosteroids analogs. Tetrahedron 2000; 56: 6171-81.

[8]   Schaefer S, Medeiro AS, Ramirez JA, *et al.* Brassinosteroid-driven enhancement of the *in vitro* multiplication rate for the marubakaido apple rootstock [*Malus prunifolia* (Willd.) Borkh]. Plant Cell Rep 2002; 20: 1093-97.

[9]   Kirk KL, Cohen LA. Photochemical decomposition of diazonium fluoroborates. Application to the synthesis of ring-fluorinated imidazoles. J Am Chem Soc 1971; *93:* 3060-61.

[10]   Lim SH, Chang SC, Lee JS, Kim SK, Kim SY. Brassinosteroids affect ethylene production in the primary roots of maize (*Zea mays* L.). J Plant Biol 2002; 45: 148-53.

[11]   Arteca RN. Brassinosteroids. In: Davies PJ, Ed. Plant hormones: Physiology, Biochemistry and Molecular Biology. Dordrecht: Kluwer Academic Publishers 1995; pp. 206-13.

[12]   Woeste KE, Vogel JP, Kieber JJ. Factors regulating ethylene biosynthesis in etiolated *Arabidopsis thaliana* seedlings. Physiol Plant 1999; 105: 478-84.

[13]   Pereira-Netto AB, Cruz-Silva CTA, Schaefer S, Ramírez JA, Galagovsky LR. Brassinosteroid-stimulated branch elongation in the Marubakaido apple rootstock. Trees-Struct Func 2006; 20: 286-91.

[14]   Neljubow D N. Uber die horizontale nutation der stengel von *Pisum sativum* und einiger anderen. Beih Bot Zentralbl 1901; 10: 128-39.

[15]   Ecker JR. The ethylene signal-transduction pathway in plants. Science 1995; 268: 667-75.

[16]   Smalle J, Haegman M, Kurepa J, Van Montagu M, van Der Straeten D. Ethylene can stimulate *Arabidopsis* hypocotyl elongation in the light. Proc Natl Acad Sci USA 1997; 94: 2756-61.

[17]   Jackson MB. Ethylene-promoted elongation: an Adaptation to submergence stress. Ann Bot 2008; 101: 229-48.

[18]   Back TG, Pharis, RP. Structure-activity studies of brassinosteroids and the search for novel analogues and mimetics with improved bioactivity. J Plant Growth Regul 2003; 22: 350-61.

[19]   Pereira-Netto AB, Schaefer S, Galagovsky LR, Ramirez JA. Brassinosteroid-driven modulation of stem elongation and apical dominance: applications in micropropagation. In: Hayat S, Ahmad A, Eds. Brassinosteroids: bioactivity and crop productivity. Dordrecht: Kluwer Academic Publishers 2003; pp 129-57.

[20]   Bishop GJ, Yokota T. Plants steroid hormones, brassinosteroids: Current highlights of molecular aspects on their synthesis/metabolism, transport, perception and response. Plant Cell Physiol 42: 114-20.

[21]   Mathur J, Molnar G, Fujioka S, *et al.* Transcription of the *Arabidopsis* CPD gene, encoding a steroidogenic cytochrome P450, is negatively controlled by brassinosteroids. Plant J 1998; 14: 593-602.

[22]   Goda H, Shimada Y, Asami T, Fujioka S, Yoshida S. Microarray analysis of brassinosteroid-regulated genes in *Arabidopsis*. Plant Physiol 2002; 130: 1319-34.

[23]   Tanaka K, Asami T, Yoshida S, Nakamura Y, Matsuo T, Okamoto S. Brassinosteroid homeostasis in *Arabidopsis* is ensured by feedback expressions of multiple genes involved in its metabolism. Plant Physiol 2005; 138: 1117-25.

[24]   Gendron JM, Wang Z-Y. Multiple mechanisms modulate brassinosteroid signaling. Curr Opin Plant Biol 2007; 10: 436-41.

[25]   Lisso J, Altmann T, Mussig C. Metabolic changes in fruits of the tomato d(x) mutant. Phytochem 2006; 67: 2232-38.

[26]   Bishop GJ, Nomura T, Yokota T, *et al.* The tomato DWARF enzyme catalyzes C-6 oxidation in brassinosteroid biosynthesis. Proc Natl Acad Sci U.S.A. 1999; 96: 1761-66.

[27]   Nomura T, Kushiro T, Yokota T, Kamiya Y, Bishop GJ, Yamaguchi S. The last reaction producing brassinolide is catalyzed by cytochrome P-450s, CYP85A3 in tomato and CYP85A2 in *Arabidopsis*. J Biol Chem 2005; 280: 17873-79.

[28]   Winter H, Huber SC. Regulation of sucrose metabolism in higher plants: localization and regulation of activity of key enzymes. CRC Crit Rev Plant Sci 2000; 19: 31-67.

[29]   Sekimata K, Han SY, Yoneyama K, Takeuchi Y, Yoshida S, Asami T. A specific and potent inhibitor of brassinosteroid biosynthesis possessing a dioxolane ring. J Agric Food Chem 2002; 50: 3486-90.

[30] Choe S, Dilkes BP, Fujioka S, Takatsuto S, Sakurai A, Feldmann KA. The DWF4 gene of *Arabidopsis* encodes a cytochrome P450 that mediates multiple 22α-hydroxylation steps in brassinosteroid biosynthesis. Plant Cell 1998; 10: 231-43.

[31] Asami, T, Min YK, Nagata N, *et al.* Characterization of brassinazole, a triazole-type brassinosteroid biosynthesis inhibitor. Plant Physiol 2000; 123: 93-100.

[32] Asami T, Mizutani M, Fujioka S, *et al.* Selective interaction of triazole derivatives with DWF4, a cytochrome p450 monooxygenase of the brassinosteroid biosynthetic pathway, correlates with brassinosteroid deficiency *in planta*. J Biol Chem 2001; 276:25687-91.

[33] Asami T, Nakano T, Nakashita H, Sekimata K, Shimada Y, Yoshida S. The influence of chemical genetics on plant science: shedding light on functions and mechanism of action of brassinosteroids using biosynthesis inhibitors. J Plant Growth Regul 2003; 22: 336-49.

[34] Yamamoto R, Fujioka S, Demura T, Takatsuto S, Yoshida S, Fukuda H. Brassinosteroid levels increase drastically prior to morphogenesis of tracheary elements. Plant Physiol 2001; 125: 556-63.

[35] Clouse SD, Langford M, McMorris TC. A Brassinosteroid-insensitive mutant in *Arabidopsis thaliana* exhibits multiple defects in growth and development. Plant Physiol 1996; 111: 671-78.

[36] Symons GM, Reid JB. Hormone levels and response during de-etiolation in pea. Planta 2003; 216: 422-31.

[37] Pereira-Netto AB, Carvalho-Oliveira MMC, Ramírez JÁ, Galagovsky LR. Shooting control in *Eucalyptus grandis* × *E. urophylla* hybrid: Comparative effects of 28-homocastasterone and a 5α-monofluoro derivative. Plant Cell Tiss Organ Cult 2006; 86: 329-35.

[38] Jones TWA. Biological activities of fluorogibberellins and interactions with unsubstituted gibberellins. Phytochem 1976; 15: 1825-27.

[39] Clouse SD. Molecular genetic studies confirm the role of brassinosteroids in plant growth and development. Plant J 1996; 10: 1-8.

[40] Chon NM, Nishikawa-Koseki N, Hirata Y, Saka H, Abe H. Effects of brassinolide on mesocotyl, coleoptile and leaf growth in rice seedlings. Plant Produc Sci 2000; 3: 360-65.

[41] Kohout L, Strnad M, Kaminek M. Types of brassinosteroids and their bioassay. In: Cutler HG, Yokota T, Adam G, Eds. Brassinosteroids, chemistry, bioactivity, and applications. Washington DC, ACS Symposium Series 474. American Chemical Society 1991; pp 56-73.

CHAPTER 4

Brassinosteroids as Mediators of Plant Biotic Stress Responses

Marcelo Lattarulo Campos and Lázaro Eustáquio Pereira Peres*

Department of Biological Sciences (LCB), Escola Superior de Agricultura "Luiz de Queiroz", University of São Paulo – USP. Av. Padua Dias, 11 CP. 09, 13418-900 Piracicaba – SP, Brazil

Abstract: From passive life forms to living organisms that can sense a myriad of external signals and alter their development, our perception of how plants interact with the environment has changed profoundly. We now know that plants can promptly respond to biotic stressors by quickly reallocating resources from growth to defensive traits. Which signals do plants use to finely tune their "growth versus defense balance"? Hormones seem to fulfill this role, since they are associated with almost every process in plant development, and also with stress responses. For this reason, plants are constantly modulating hormonal pathways in order to better allocate internal resources. Here, we discuss how brassinosteroids (BRs), steroidal plant hormones known to be potent growth regulators, work as strong mediators of plant biotic stress responses. Interaction between BRs and other stress hormones, like jasmonates and salicylic acid, important to build-up defensive barriers necessary to cope with insects and microbes are also discussed. Finally, we present evidence that these plant steroids are not only directly involved in defense responses against pest and pathogens, but they are also key regulators in the resource allocation decision.

Keywords: Insect herbivory, jasmonates, resource allocation, salicylic acid, systemic acquired resistance (SAR).

1. TO GROW OR TO DEFEND: A PLANT DILEMMA

In order to avoid certain stressful situations, many living organisms can simply generate movement: a prey, for example, can try to escape from its predator just by outrunning it. As sessile organisms, plants have long been considered passive life forms in which developmental processes follow predetermined programs so that stressful conditions lead to reductions in fitness. If this is true, how can a world with plants constantly facing a myriad of stressors like pests and pathogens, water and nutrient deprivation, temperature changes, and so forth, still be green [1]?

Our view of plants as passive organisms that simply endure its surroundings has changed as we now know that plants can recognize a plethora of environmental signals and alter their development in order to respond to stressful conditions. For example, upon insect herbivory, many plants can trigger the production of specific volatiles which are capable of attracting parasites or predators that are natural enemies of the herbivorous insect [2,3], thus reducing the damage caused by it.

One of the best ways to observe how plastic plant development is and how plants can respond to external stimuli and alter their physiology is to focus on biotic stresses, *i.e.* the stresses caused by other living organisms. Observing plant interactions with herbivorous insects, Ehrlich & Raven [4] proposed the co-evolution theory, where plants develop innumerable defensive barriers to fend themselves against insects, such as the production of toxic chemical compounds [5]. Insects concomitantly develop strategies to overcome these barriers, like the production of detoxification compounds [6]. As proposed by Ehrlich & Raven [4], this evolutionary arms race is not only restricted to plant-insect interactions, but to any two groups of organisms with a close and evident ecological relationship.

The production of defensive barriers is nevertheless costly. Some defensive traits like trichomes and cuticle can be constitutively present in plants, while other specific resistance barriers are only expressed

---

*****Address correspondence to Lázaro Eustáquio Pereira Peres:** Department of Biological Sciences (LCB), Escola Superior de Agricultura "Luiz de Queiroz", University of São Paulo – USP. Av. Padua Dias, 11 CP. 09, 13418-900 Piracicaba – SP, Brazil; E-mail: lazaropp@esalq.usp.br

facultatively, when needed. Individuals that can optimize this tradeoff and allocate resources into growth and reproduction, when not under attack, will thus have a fitness advantage [7, 8]. The reallocation of resources from growth to defense creates an internal dilemma for plants: how much energy should be spent (and for how long) in defense, in detriment of growth and reproduction or vice-versa? If plants have to balance their resource allocation in the face of this dilemma, then extensive cross-talk between plant defense signaling and growth/development must exist.

Here, we discuss new evidences that brassinosteroids (BR), hormones known to be potent regulators of plant growth and development [9], can also work as signals controlling resource allocation decision during attack of pests and pathogens, thus serving as mediators of plant biotic stress responses.

2. HORMONES: MOLECULES MODULATING PLANT RESOURCE ALLOCATION DECISION

From seed germination to gravitropism, apical dominance, bud formation, light perception, flowering set, fruit maturation, leaf senescence and many other processes, hormones are molecules associated with almost every step of plant development. Most hormones are also involved in modulating biotic stress responses, and, plants can finely control metabolism (synthesis/degradation), translocation, and signal transduction of hormones, in order to optimize their internal resource allocation [10].

Some hormones are well known for modulating growth processes and plant's defenses against biotic stresses. Three of these hormones are of paramount importance for their dual task in this plant growth versus plant defense dillema: Jasmonates (JA), salicylic acid (SA) and ethylene.

Initially known for their role in senescence, inhibition of root growth and development of reproductive structures [11-14], JAs are now under the spotlight for their function in defense against biotic stress. It has been demonstrated, for example, that both, insect herbivory and attack by some pathogens can strongly stimulate the biosynthesis of these ubiquitously occurring oxylipin-derived hormones [15-17].

Ethylene seems to work similarly to JAs while modulating response to biotic stress. This plant hormone, primarily involved with developmental processes such as fruit ripening and, leaf and flower senescence [18], is also involved in defense against pests and pathogens. Upon insect attack, for example, an ethylene burst triggered in plants induces the expression of anti-herbivory proteins like polyphenol oxidase and peroxidases, and also the synthesis of defensive metabolites such as glucosinolates [19]. Interestingly, ethylene and JA pathways positive cross-talk enhances the defenses against biotic stresses [20]. Salicylic acid (SA), on the other hand, seems to share negative cross-talk with both, JA and ethylene (please see section 4.2, below). SA is involved in processes like seed germination and stomatal closure, but it is better known for its involvement in defense response against biotrophic pathogens and viruses [21]. Classical experiments showed that SA application could enhance resistance against pathogens such as the tobacco mosaic virus [21]. SA is the main signaling component in the so called systemic acquired resistance (SAR) mechanism, in which infection by a single microbial pathogen triggers resistance throughout the whole plant [22, 23]. Transgenic plants unable to accumulate SA fail to establish SAR and can not prevent pathogen spread [24].

Different from JA, SA and ethylene, BRs are better known for their role as powerful regulators of growth processes [9]. BRs have been demonstrated to control cell division and differentiation, stem elongation, leaf development, flowering and xylem formation [25]. In the late 90s, Szekeres *et al.,* [26] provided the first evidence that BRs are also involved in the modulation of plant resource allocation. They observed that the overexpression of a steroid hydroxylase involved in BR biosynthesis in *Arabidopsis* led to the induction of genes involved in pathogen defense, like *PR1* and *PR2*. Substantial support to the hypothesis that BRs are involved in the defense against biotic stress appeared later. However, in order to understand how BRs are related to plant defense responses against other organisms, we should look at systemin, a polypeptide capable of inducing accumulation of anti-herbivore proteins in tomato [27], first.

3. THE SR160/BRI1 STORY

Upon mechanical wounding or insect herbivory, some plants like tomato and potato are able to accumulate an anti-digestive class of proteins known as serine proteinase inhibitors (PIs). PIs act in plant defense by interacting with proteins present in the digestive tract of the attacking insects, probably reducing digestibility and thus the amount of nutrients absorbed [28]. Remarkably, not only the wounded leaves produce PIs, but also non-wounded leaves, a process known as systemic response [29]. The systemic response probably has an ecological role by defending plants from further attack by pests. In order to identify a signal capable of inducing the systemic response in tomato, Pearce and co-workers [27] isolated an 18-amino acid peptide called systemin, which, when applied at very low concentrations (10^{-15} M), was able to induce accumulation of PIs in tomato leaves. Systemin apparently works *via* the activation of the octadecanoid pathway, leading to a burst of JAs, which stimulates the expression of PI's. Not surprisingly, mutants impaired in JA synthesis are defective in PIs production upon systemin feeding [30].

In order to identify the putative systemin receptor, Scheer & Ryan [31] used suspension-cultured *Solanum peruvianum* cells and ^{125}I-labeled systemin to specifically isolate a 160kD leucine-rich repeat receptor-like kinase (LRR-RLK) named SR160. A few years later, Montoya and co-workers [32] cloned the tomato BR receptor, a homolog of the *Arabidopsis* LRR-RLK Brassinosteroid Insensitive 1 (BRI1), and surprisingly found that the sequences from SR160 and the tomato BRI1 were essentially identical. Those findings suggested that the SR160 and BRI1 are actually one receptor with dual ligands: the defense-related polypeptide systemin and the growth-related hormone BR.

How can two different ligands (systemin and BR) with very distinct functions, converge to bind to a single receptor, SR160/BRI1 [33]? The fact that BRs are potent growth regulators and systemin has a clear role in plant defense would strongly suggest that SR160/BRI1 has a role in "resolving" the plant dilemma by acting as a hub in cross-talk between biotic stress responses versus growth-related signals. One could imagine a model where systemin and BRs compete for binding to the SR160/BRI1. Under optimal growth conditions (no biotic stress), no systemin is produced and BRs can bind to the receptor and trigger growth-related responses. However, upon insect herbivory, a burst of systemin would create an environment where the concentration of systemin would be much higher when compared to the concentration of BRs. Then, systemin would be able to outcompete BRs to reach the receptor, thus triggering defense-related responses. The dimerization of receptor kinases upon binding to its ligands seems to be necessary to induce downstream responses in plants [34, 35]. It is now well established that upon BR binding, BRI1 can heterodimerize with another LRR-RLK, the BRI1-associated receptor kinase 1 (BAK1, also named SERK3 - Somatic Embryogenesis Receptor Kinase 3). So, does the binding to systemin trigger a different type of dimerization and consequently a different downstream response? A twist in the dual ligand SR160/BRI1 story happened recently, when two groups, independently, showed that the tomato null mutant for BRI1, *curl-3,* exhibits a normal response to systemin [36, 37]. Later on, using fluorescent-labeled systemin, Malinovski and co-workers [38] showed that systemin was able to bind to BRI1 without triggering any downstream signal. These results strongly suggest that: 1) BRI1 is a systemin binding protein, but *not* its physiological receptor and; 2) the putative systemin receptor probably shares similar properties with BRI1 [39].

Interestingly, the latest evidences point now to a new direction for the role of BRs in the resource allocation decision dilemma, at a different level than the BRI1 receptor.

4. BRASSINOSTEROIDS: DUAL ROLE HORMONES MODULATING GROWTH VERSUS DEFENSE RESPONSE

4.1. Brassinosteroids x Jasmonates Interaction

Systemin triggers the production of PIs most likely through the activation of the JA pathway. Although it is very unlikely that BRI1 is the receptor for both systemin and BRs, recent evidence suggest that cross-talk between JA and BR pathways are involved in the generation of anti-herbivory traits.

It is well known that tomato plants impaired in biosynthesis or sensing of JAs, like the *jasmonic acid insensitive 1-1 (jai1-1)* mutant, are defective in several defense responses, such as trichome formation, accumulation of some secondary metabolites and expression of PIs, upon wounding [12]. Surprisingly, work in our laboratory recently showed that the *dumpy (dpy),* a tomato mutant impaired in BRs biosynthesis, presents higher trichome density in the leaves, higher levels of secondary metabolites and higher expression of PIs upon mechanical wounding [40]. In order to further investigate the unexpected defense-related features of the *dpy* mutant, we challenged the mutant with the leaf-miner (*Tuta absoluta*), the main tomato pest in its center of origin, South America. The mortality of larvae growing on the *dpy* plants was shown to be higher when compared to the mortality rate of larvae grown on both, wild type and *jai1-1* plants. Moreover, our double mutant between *dpy* and *jai1-1*, the *dpy* x *jai1-1*, still presented traits characteristic of a BR impaired mutant, such as severe dwarfism and darker green and curled leaves (Fig. **1A**). On the other hand, all of the overproduced defensive traits found for the *dpy* mutant were reduced, in the double mutant, to the levels observed in *jai1-1*, which means that the *dpy* x *jai1-1* plants were defective in trichome formation (Fig. **1B-D**), had lower levels of secondary metabolites and reduced PI production. *T. absoluta* larvae mortality was also reduced in *dpy x jai1-1* to levels similar of the *jai1-1* plants [40]. The conclusions that can be drawn from these data are that: 1) as expected, JAs control the production of many anti-herbivory traits in tomato, and; 2) BRs can negatively control the production of defensive barriers by acting on the JA pathway (Fig. **1E**).

The hypothesis that BRs negatively interacts with the JA pathway, can be extended to other plant species. The *Arabidopsis* gene *CORONATINE INSENSITIVE1* (*COI1*) encodes an F-box protein that plays a central role in JA signaling, acting as the JA receptor [14]. The fact that JA is capable of inhibiting root growth is normally used as a bioassay to evaluate JA-sensitive mutants and, when *coi1* mutant plants are grown in the presence of JA, almost no root growth inhibition is observed. Based on the JA root growth phenotype, Ren and co-workers [41] isolated suppressors of the *coi1* mutant (*i.e. coi1* plants that present root inhibition in the presence of JA). Interestingly, one of the isolated suppressors was *psc1,* a mutant that harbors a leaky mutation in the *DWARF4* gene, which encodes an enzyme involved in BR synthesis [42]. This data indicates that in the absence (or low levels) of BRs, the sensibility to JA is increased. Thus, the reduced BR synthesis in *psc1* diminishes the negative effect of BRs on JA sensibility (as observed in the root growth assay) which reverses the *coi1* phenotype. Finally, Yang and co-workers [43] observed that silencing BAK1, a key component of the BR signaling pathway [44, 45] (please see section 4.3) leads to high levels of PI accumulation upon application of methyl-JA in *Nicotiana attenuata*.

Taken together, these data strongly suggest that a reduction in BR sensitivity (or endogenous levels) can cause an enhancement on JA response, which favors the reallocation of resources from growth to defense. In the case of tomato, a smart way to stimulate the reallocation from growth to defense would be a reduction in BR sensitivity, through the production of systemin, which might down-regulate the BR signaling through the competition for the BR receptor SR160/BRI1, as discussed in the previous section.

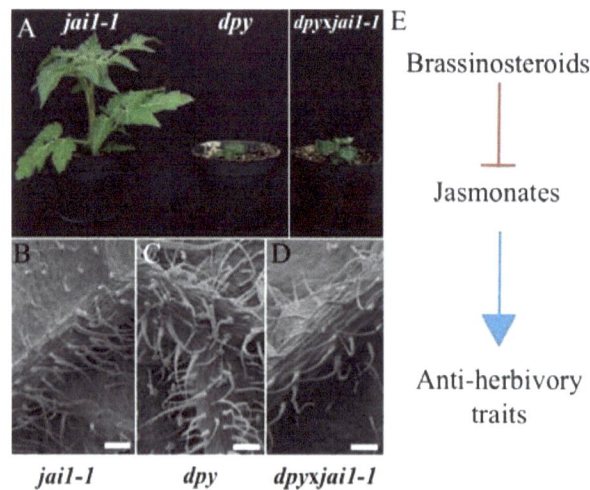

Figure 1: The cross-talk between jasmonates and brassinosteroids pathways in the formation of anti-herbivory traits in tomato (according to ref. 40). When crossing the jasmonate insensitive mutant *jai1-1* with the brassinosteroid deficient *dpy,* the resulting double mutant presents a phenotype defective in BR production, severe dwarfism and dark curled leaves (A). However, defensive traits like the presence of trichomes, which are produced in large amounts in *dpy* (C), are similar to *jai1-1* (B) in the *dpyxjai1-1* double mutant (D). The conclusion that can be drawn is that BRs negatively interacts with the JA pathway in order to control the formation of anti-herbivory traits in tomato (E). Bars = 200μm.

4.2. Brassinosteroids and Salicylic Acid: Additive Roles in Plant Pathogen Defense

Concerning the resistance against biotic stresses, the SA and JA pathways seem to cross-talk in a complex way, which can be described, generally, as antagonistic. SA can, for example, inhibit systemin and JA-induced synthesis of PIs [46], and also block the expression of JA responsive genes like *PDF1.2* and *VSP* [47]. JA can also suppress the production of SA [48]. Some bacteria use the SA x JA antagonism to hijack plant defensive barriers by producing a substance with a chemical structure similar to JA, the phytotoxin coronatine, thus suppressing the SA-related defense system [49]. Insects use a similar strategy, inducing salicylic acid defenses to further suppress JA defensive barriers [50].

As already proposed in section 4.1, BRs seem to negatively interact with the JA pathway, blocking the formation of JA-related defensive traits, in a mechanism that somehow resembles the effect of SA in the JA pathway. This begs the question, is there any relationship between BR and the SA pathway? Do these hormones interact with each other in order to generate defenses against microbial pathogens? The answer is not known yet, but it is likely to be positive.

BRs have been demonstrated to induce disease resistance to virus, bacteria and fungus, when applied to leaves of tobacco and rice. Surprisingly, this BR-induced resistance is observed even in *nahG* transgenic plants, plants unable to accumulate SA, suggesting that BR induces a defensive pathway that is independent of SA [51]. However, it was also observed that simultaneous activation of the BR-induced and SA-induced resistance led to an additive effect on disease resistance [51]. Thus, although independent, the BR-induced and SA-induced resistance pathways can exhibit additive protective effect against some biotic stress.

4.3. The Dual Role of BAK1

There are lines of evidence pointing to the fact that components of the BR signaling pathway could orchestrate defensive responses, even in a BR-independent way. The leucine-rich repeat receptor-like kinase LRR-RLK BAK1/SERK3 plays an essential role in BR signaling by physically interacting with the BR receptor BRI1, forming a larger BR receptor complex [44, 45]. The association between BRI1 and BAK1, forming a receptor kinase pair, can be genetically observed through mutant phenotypes. The overexpression of BAK1 leads to a BRI1-like overexpression phenotype, *i.e.*, plants more sensitive to BRs, displaying narrower and longer leaves, with larger petioles. Conversely, the *bak1* knockout mutation produces dwarf plants, with dark green curled leaves, a phenotype resembling the *bri1* mutant [52].

Two groups have demonstrated the importance of BAK1 for plant responses against pathogens [34, 53]. Plants can sense potential pathogen invaders through the recognition of pathogen-associated molecular patterns (PAMPs), typical (and slowly evolving) molecular structures shared by many microbial organisms. Flagellin and the elongation factor TU are examples of PAMPs, which are recognized by plant transmembrane pattern recognition receptors (PRR), triggering defensive responses like the production of reactive oxygen species, expression of PR1 and salicylic acid production [54]. This process is known as PAMP-triggered immunity [55].

The LRR-RLK FLAGELLIN-INSENSITIVE 2 (FLS2) is the PRR that recognizes the bacterial elicitor flaggelin [56]. For their inability to recognize flaggelin, *fls2* mutants are more sensitive to pathogens like *Pseudomonas syringae* [54]. Looking for *Arabidopsis* insertional mutants with different responsiveness to flagellin, Chinchilla *et al.* [34] isolated two mutants with reduced sensitivity to flagellin. Both of them were later found to be null mutants of the *BAK1* gene. In addition, Kemmerling and co-workers [53] found increased *BAK1* transcript accumulation in plants of *Arabidopsis* infected by many *P. syringae* strains. Furthermore, FLS2 was shown to dimerizes with BAK1 within the first minutes of stimulation with flagellin [34], a mechanism that resembles BRI1-BAK1 complex formation. Interestingly, *bak1* plants present normal flagellin bidding but abnormal early and late flagellin-triggered responses, which corroborate the positive role of LRR-RLK BAK1 in pathogen defense signaling.

BAK1 is thus a ligand with dual role: it can interact with both, the BR receptor BRI1, triggering growth responses, and with FLS2, triggering defense responses against pathogen attack. The dual role of BAK1 raises an intuitive question: do BRs exert any role in the FLS2-BAK1 function or responses? The available evidences suggest that they do not. First, BRs are not necessary for FLS2-flagellin binding [34]. Second, BR-deficient mutants do not present altered response to flagellin [53]. These results show that BAK1, a BR-signaling pathway component, acts independently of BRs in order to regulate defense responses to biotic stress. But what about the opposite? Can flagellin changes the way BAK1 interacts with BRI1? If that is the case, recognition of pathogens like *P. syringae* can, for example, reduce the formation of the BRI1-BAK1 complex and consequently reduce the responsiveness of the plant to BRs. Although no data are yet available, it is possible to speculate that this mechanism might be a way to reallocate resources from growth to pathogen defense upon microbial attack.

5. A NEW "GREEN REVOLUTION"?

The utilization of mutants with altered synthesis/response to gibberellins, was one of the bases of the "green revolution" in the late 60's [57, 58]. Interestingly, the shorter and stiff-strawed, high-yield green revolution crops were also more resistant to some abiotic stresses, like rain and wind [57].

In a world looking for sustainability, the permanent use of pesticides has solved many problems but has also created various problems such as environmental damage, lower productivity, increase in the production cost, human health problems, and many others [59]. Thus, should we start looking at hormones again in order to develop plants more resistant to pests and pathogens? Can we think about a new "green revolution" by developing crop varieties with reduced or no requirement for pesticides application, thus leading to less environmental harm?

We have shown in this chapter how BRs, plant hormones known to be potent growth regulators, can also exert a fundamental role in the control of resistance to biotic stress. BR deficient mutants usually present dwarf phenotypes, somehow resembling the gibberellin mutants used in the "green revolution". Sakamoto and co-workers [60], for example, described a dwarf rice BR-deficient mutant with increased biomass production and higher grain yield. The evidences reviewed in this chapter suggest that modulation of BR production/sensitivity can induce resistance against pests or pathogens, thus allowing a reduction in both, the amount of pesticides used in agriculture, and their environmental harm.

LIST OF ABBREVIATIONS

BAK1   =   BRI1-Associated receptor Kinase 1

BRs   =   Brassinosteroids

BRI1   =   Brassinosteroid Insensitive 1

coi1   =   *coronatine insensitive 1*

dpy   =   *dumpy*

FLS2   =   Flagellin-inSensitive 2

JAs   =   Jasmonates

jai1-1   =   *jasmonic acid insensitive 1-1*

LRR-RLK =   Leucine-Rich Repeat Receptor-Like Kinase

PAMPs   =   Pathogen-Associated Molecular Patterns

PI   =   Serine Proteinase Inhibitor

PRR   =   Pattern Recognition Receptors

SA   =   Salicylic Acid

SAR   =   Systemic Acquired Resistance

REFERENCES

[1]   Boller T, He SY. Innate immunity in plants: an arms race between pattern recognition receptors in plants and effectors in microbial pathogens. Science 2009; 324: 742-4.

[2]   Birkett MA, Campbell CAM, Chamberlain K, *et al.* New roles for cis-jasmonate as an insect semiochemical and in plant defense. Proc Natl Acad Sci USA 2000; 97: 9329-34.

[3]   Paré PW, Tumlinson JH. Plant volatiles as a defense against insect herbivores. Plant Physiol 1999; 121: 325-32.

[4]   Ehrlich PR, Raven PH. Butterflies and plants: A study in coevolution. Evolution 1964; 18: 586-608.

[5]   Kliebenstein DJ, Kroymann J, Brown P, *et al.* Genetic control of natural variation in Arabidopsis glucosinolates accumulation. Plant Physiol 2001; 126: 811-25.

[6]   Schuler MA. The role of cytochrome P450 monooxygenases in plant-insect interactions. Plant Physiol 1996; 112: 1411-9.

[7]   Baldwin IT. Jasmonate-induced responses are costly but benefit plants under attack in native populations. Proc Natl Acad Sci USA 1998; 95: 8113-8.

[8]   Karban R, Agrawal AA, Mangel M. The benefits of induced defenses against herbivores. Ecology 1997; 78: 1351-5.

[9]   Clouse SD, Sasse, J. Brassinosteroids: Essential regulators of plant growth and development. Annu Rev Plant Phys 1998; 49: 427-51.

[10]   Grant MR, Jones JD. Hormone (dis)harmony moulds plant health and disease. Science 2009; 324: 750-2.

[11]   He Y, Fukushige H, Hilderbrand DF, Susheng G. Evidence supporting a role of jasmonic acid in Arabidopsis leaf senescence. Plant Physiol 2002; 128: 876-84.

[12]   Li L, Zhao Y, McCaig BC, *et al.* The tomato homolog of *CORONATINE-INSENSITIVE1* is required for the maternal control of seed maturation, jasmonate-signaled defense responses, and glandular trichome development. Plant Cell 2004; 16: 126-43.

[13]   Xie DX, Feys BF, James S, Nieto-Rostro M, Turner JG. *COI1:* An *Arabidopsis* gene required for jasmonate-regulated defense and fertility. Science 1998; 280: 1091-4.

[14]   Yan J, Zhang C, Gu M, *et al.* The *Arabidopsis CORONATINE INSENSITIVE1* protein is a jasmonate receptor. Plant Cell 2009; 21: 2220–36.

[15]   McConn M, Creelman RA, Bell E, Mullet JE, Browse J. Jasmonate is essential for insect defense in *Arabidopsis.* Proc Natl Acad Sci USA 1997; 94: 5473-7.

[16]   Reymond P, Weber H, Damond M, Farmer EE. Differential gene expression in response to mechanical wounding and insect feeding in *Arabidopsis.* Plant Cell 2000; 12: 707-20.

[17]   Thomma BPHJ, Eggermont K, Penninckx IAMA, Mauch *et al.* Separate jasmonate-dependent and salicylate-dependent defense-response pathways in *Arabidopsis* are essential for resistance to distinct microbial pathogens. Proc Natl Acad Sci USA 1998; 95: 15107-11.

[18]   Yang Y, Wu Y, Pirrello J, *et al.* Silencing *Sl-EBF1* and *Sl-EBF2* expression causes constitutive ethylene response phenotype, accelerated plant senescence, and fruit ripening in tomato. J Exp Bot 2010; 61: 697-708.

[19]   Dahl CC, Baldwin IT. Deciphering the role of ethylene in plant-herbivore interactions. J Plant Growth Regul 2007; 26: 201-9.

[20]   Onkokesung N, Gális I, Dahl CC, Matsuoka K, Saluz HP, Baldwin IT. Jasmonic acid and ethylene modulate local responses to wounding and simulated herbivory in *Nicotiana attenuata* leaves. Plant Physiol 2010; 153: 785-98.

[21]   Durner J, Shah J, Klessig DF. Salicylic acid and disease resistance in plants. Trends Plant Sci 1997; 2: 266-74.

[22]   Durrant WE, Dong X. Systemic acquired resistance. Annu Rev Phytopathol 2004; 42: 185-209.

[23]   Vlot AC, Dempsey DA, Klessig DF. Salicylic acid, a multifaceted hormone to combat disease. Annu Rev Phytopathol 2009; 47: 177-206.

[24]   Gaffney T, Friedrich L, Vernooij B, *et al.* Requirement of salicylic acid for the induction of systemic acquired resistance. Science 1993; 261: 754-6.

[25]   Zhang S, Wei Y, Lu Y, Wang X. Mechanisms of brassinosteroids interacting with multiple hormones. Plant Signal Behav 2009; **4:** 1117-20.

[26]   Szekeres M, Németh K, Koncz-Kálmán Z, *et al.* Brassinosteroids rescue the deficiency CYP90, a cytochrome P450 controlling cell elongation and de-etiolation in *Arabidopsis.* Cell 1996; 85: 171-82.

[27]   Pearce G, Strydom D, Johnson S, Ryan CA. A polypeptide from tomato leaves induces wound-inducible proteinase inhibitor proteins. Science 1991; 253: 895-8.

[28] Ryan CA. Protease inhibitors in plants: Genes for improving defenses against insects and pathogens. Annu Rev Phytopathol 1990; 28: 425-49.

[29] Green TR, Ryan CA. Wound-induced proteinase inhibitors in plants leaves: a possible defense mechanism against insects. Science 1972; 175: 776-7.

[30] Stratmann JW. Long distance run in the wound response – jasmonic acid is pulling ahead. Trends Plant Sci 2003; 8: 247-50.

[31] Scheer JM, Ryan CA. A 160-kD systemin receptor on the surface of *Lycopersicon peruvianum* suspention-cultured cells. Plant Cell 1999; 11: 1525-36.

[32] Montoya T, Nomura T, Farrar K, Kaneta T, Yokota T, Bishop GJ. Cloning the tomato *Curl3* gene highlights the putative dual role of the leucine-rich repeat receptor kinase tBRI1/SR160 in plant steroid and peptide hormone signaling. Plant Cell 2002; 12: 3163-76.

[33] Szekeres M. Brassinosteroid and systemin: two hormones perceived by the same receptor. Trends Plant Sci 2003; 8: 102-4.

[34] Chichilla D, Zipfel C, Robatzek S, *et al.* A flagellin-induced complex of receptor FLS2 and BAK1 initiates plant defence. Nature 2007; 448: 497-501.

[35] Torii KU (2000). Receptor kinase activation and signal transduction in plants: and emerging picture. Curr Opin Plant Biol 2000; 3: 361-7.

[36] Holton N, Cano-Delgado A, Harrison K, Montoya T, Chory J, Bishop GJ. Tomato BRASSINOSTEROID INSENSITIVE1 is required for systemin-induced root elongation in *Solanum pimpinellifolium* but is not essential for wound signaling. Plant Cell 2007; 19: 1709-17.

[37] Lanfermeijer FC, Staal M, Malinowski R, Stratmann JW, Elzenga JTM. Micro-electrode flux estimation confirms that the *Solanum pimpinellifolium cu3* mutant still responds to systemin. Plant Physiol 2008; 146: 129-39.

[38] Malinowski R, Higgins R, Luo Y, *et al.* The tomato brassinosteroid receptor BRI1 increases binding of systemin to tobacco plasma membranes, but is not involved in systemin signaling. Plant Mol Biol 2009, 70: 603-16.

[39] Hind SR, Malinowski R, Yalamanchili R, Stratmann JW. Tissue-type specific systemin perception and the elusive systemin receptor. Plant Signal Behav 2010; 5: 42-4.

[40] Campos ML, Almeida M, Rossi ML, *et al.* Brassinosteroids interact negatively with jasmonates in the formation of anti-herbivory traits in tomato. J Exp Bot 2009; 60: 4347-61.

[41] Ren C, Han C, Peng W, *et al.* A leaky mutation in *DWARF4* reveals an antagonistic role of brassinosteroid in the inhibition of root growth by jasmonate in *Arabidopsis*. Plant Physiol 2009; 151: 1412-20.

[42] Choe S, Dilkes BP, Fujioka S, Takatsuto S, Sakurai A, Feldmann KA. The *DWF4* gene of Arabidopsis encodes a cytochrome P450 that mediates multiple 22α-hydroxylation steps in brassinosteroid biosynthesis. Plant Cell 1998; 10: 231-43.

[43] Yang DH, Hettenhausen C, Baldwin IT, Wu J. BAK1 regulates the accumulation of jasmonic acid and the levels of trypsin proteinase inhibitors in *Nicotiana attenuata*'s responses to herbivory. J Exp Bot 2011; 62: 641-52.

[44] Li J, Wen J, Lease KA, Doke JT, Tax FE, Walker JC. BAK1, an *Arabidopsis* LRR receptor-like protein kinase, interacts with BRI1 and modulates brassinosteroid signaling. Cell 2002; 110: 213-22.

[45] Russinova E, Borst JW, Kwaaitaal M, *et al.* Heterodimerization and endocytosis of Arabidopsis brassinosteroid receptors BRI1 and AtSERK3 (BAK1). Plant Cell 2004; 16: 3216-29.

[46] Doares SH, Narvaez-Vasquez J, Conconi A, Ryan CA. Salicylic acid inhibits synthesis of proteinase inhibitors in tomato leaves induced by systemin and jasmonic acid. Plant Physiol 1995; 108: 1741-6.

[47] Spoel SH, Koornneef A, Claessens SMC, *et al.* NPR1 modulates cross-talk between salycilate- and jasmonate-dependent defense pathways through a novel function in the cytosol. The Plant Cell 2003; 15: 760-70.

[48] Diezel C, Dahl CC, Gaquerel E, Baldwin. Different lepidopteran elicitors account for cross-talk in herbivory-induced phythohormone signaling. Plant Physiol 2009; 150: 1576-86.

[49] Brooks DM, Bender CL, Kunkel BN. The *Pseudomonas syringae* phytotoxin coronatine promotes virulence by overcoming salicylic acid-dependent defenses in *Arabidopsis thaliana*. Mol Plant Pathol 2005; 6: 629-39.

[50] Zarate SI, Kempema LA, Walling LL. Silverleaf whitefly induces salicylic acid defenses and suppresses effectual jasmonic acid defenses. Plant Physiol 2007; 143: 866-75.

[51] Nakashita H, Yasuda M, Nitta T, *et al.* Brassinosteroid function in a broad range of disease resistance in tobacco and rice. Plant J 2003; 33: 887-98.

[52] Nam KH, Li J. BRI1/BAK1, a receptor kinase pair mediating brassinosteroid signaling. Cell 2002; 110: 203-12.

[53] Kemmerling B, Schwedt A, Rodriguez P, *et al.* The BRI1-associated kinase 1, BAK1, has a brassinolide-independent role in plant cell-death control. Curr Biol 2007; 17: 1116-22.

[54] Zipfel C, Robatzek S, Navarro L, *et al.* Baterial disease resistance in *Arabidopsis* through flagellin perception. Nature 2004; 428: 764-7.

[55] Jones JDG, Dangl JL. The plant immune system. Nature 2006; 444: 323-9.

[56] Gómez-Gómez L, Boller T. FLS2: an LRR receptor-like kinase involved in the perception of the bacterial elicitor flagellin in *Arabidopsis*. Mol Cell 2000; 5: 1003-11.

[57] Peng J, Richards DE, Hartley NM, *et al.* "Green revolution" genes encode mutant gibberellin response modulators. Nature 1999; 400: 256-61.

[58] Sasaki A, Ashikari M, Ueguchi-Tanaka M, *et al.* A mutant gibberellin-synthesis gene in rice. Nature 2002; 416: 701-2.

[59] Wilson C, Tisdell C. Why farmers continue to use pesticides despite environmental, health and sustainability costs. Ecol Econom 2001; 39: 449-62.

[60] Sakamoto T, Morinaka Y, Ohnishi T, *et al.* Erect leaves caused by brassinosteroid deficiency increase biomass production and grain yield in rice. Nat Biotechnol 2006; 24: 105-9.

CHAPTER 5

Enhanced Tolerance to Heavy Metals

Mohammad Yusuf, Shamsul Hayat[*], Qazi Fariduddin and Aqil Ahmad

Plant Physiology Section, Department of Botany, Aligarh Muslim University, Aligarh-202 002, India

Abstract: The studies discussed here include the effect of brassinosteroids (BRs) on seed germination, growth, and development, yield characteristics, nitrogen metabolism, including nitrate reductase activity, chlorophyll content, photosynthesis and antioxidant system of plants grown under different levels of heavy metals. This chapter is expected to provide an easy access to the pivotal role of BRs on heavy metal detoxification.

Keywords: 24-epibrassinolide (EBL), 28-homobrassinolide (HBL), heavy metal detoxification, heavy metal stress.

INTRODUCTION

Before 1970, it was assumed that the main plant growth processes were regulated by five types of classical phytohormones, namely auxins, cytokinins, gibberellin, abscisic acid and ethylene. The steroids, as hormones, were known only in animals, but they have now been discovered in plants too. The identification of plant steroidal hormones was the result of nearly 30 years of efforts to identify this novel growth promoting substances, first in pollen [1]. Early studies by J.W. Mitchell and co-workers showed that the strong growth stimulating activity was found in the organic solvent extract of the pollen from the rape plant (*Brassica napus* L.). These, then unidentified active compounds, were named brassins [2] and the bioactive compound was later named brassinolide (BL) [3]. Afterwards, many other plant steroids with structure and function similar to BL were identified throughout the plant kingdom. Now, these compounds are recognized as a new class of plant hormones named brassinosteroids (BRs), and about 70 of them have been isolated from plants [4]. Physiological studies have demonstrated that BRs can induce diverse physiological responses such as stem elongation, pollen tube growth, leaf bending and epinasty, inhibition of root growth, induction of ethylene biosynthesis, fruit ripening and xylem differentiation [5-7].

The identification of *Arabidopsis* BR biosynthetic mutants demonstrated that BRs are essential for normal plant growth and development [8, 9]. In addition, BRs have been shown to be able to confer resistance to plants against various abiotic stresses, including heavy metals [10-13, 29]. Recent studies aimed at understanding how BRs modulate stress responses, suggest that complex molecular changes underlie BR-induced stress tolerance in plants. Analyses of these changes should generate exciting results in the future and clarify whether the BRs could improve plant resistance to a range of stresses, including heavy metals. The explanation deciphering the complex interactions of BRs with other plant hormones as ABA, IAA, and GA [40, 41]. The present chapter provides an easy access to a comprehensive coverage of BRs in response to heavy metal stresses in plants.

1. BRs MECHANISM OF ACTION, RECEPTION AND TRANSPORT

The mechanism of action of BRs has been an attractive target for researchers since the elucidation of the structure of BL. Although this problem is still rather far from its final solution, many important data have brought us closer to the understanding of the mode of regulatory action of BRs.

Considering the broad range of physiological effects of BRs, it is believed that more than one molecular mechanisms of BRs action co-exist. It is very well documented that steroids function as a signaling

*Address correspondence to Shamsul Hayat: Plant Physiology Section, Department of Botany, Aligarh Muslim University, Aligarh-202 002, India; E-mail: hayat_68@yahoo.co.in

molecules in both, animals and plants. While animal steroidal hormones are perceived by a family of nuclear transcription factors receptors, BRs in plants are perceived by a cell surface receptor kinase, BRI1.

Brassinolide (BL), the most active brassinosteroid, binds to the extracellular domain of the BRI1 receptor in the plasma membrane. BRI1 is a plasma membrane localized lecuine-rich repeat (LRR)-receptor serine/threonine (S/T) kinase. The LRR-receptor kinases constitute the largest receptor class predicted in the *Arabidopsis* genome, with over 230 family members. This family has a conserved domain structure, composed of an N-terminal extracellular domain with multiple tandem (adjacent) LRR motifs, a single trans-membrane domain, and a cytoplasmic kinase domain with specificity towards serine and threonine residues. In the case of BRI1, the number of LRRs is 25. BRI1 also has a unique feature that is required for BR binding, a stretch of amino acids called the island domain that interrupts the LRRs between LRRs 21 and 22 [14]. This domain plus the flanking LRR22 compose the minimum binding site for BRs. BL binding to BRI1 triggers the interaction between BRI1 and BAK1. BRI1 is phosphorylated at multiple sites along with its intracellular domain, some of which have been shown to regulate the receptor activity. The BL signal is then transmitted to the cytoplasm by an unknown mechanism where it inhibits B1N2, which is a negative regulator of the BR biosynthetic pathway. BIN2 is a protein kinase that interacts with and phosphorylates two nearly identical transcription factors, BES1 and BZR1, negatively regulating their activities. BSU1 dephosphorylates BES1 and BZR1 to counteract the effect of BIN2. BRs regulate the expression of hundreds of genes. A significant portion of the unregulated genes is predicted to play a role in growth processes. BES1 binding activity and expression level of its target genes are enhanced synergistically by B1M1. B1M1 is another transcription factor that dimerizes with BES1 and increases its activity. Genes that are down-regulated by BR include several BR biosynthetic genes. BZR1 binds to specific elements in their promoters to repress their activity. The genes repression by B2R1 represents a negative feedback loop for the regulation of growth by BR.

2. PHYSIOLOGICAL ACTION OF BRS IN PLANTS UNDER HEAVY METAL STRESS

Review articles on (BRs) that have appeared [5] include two books [15, 16] and various articles which focus on particular aspects of BRs structure and activity [4, 17-21]. In the recent past, BRs emerged as unique steroidal phytohormones widely distributed throughout the plant kingdom. Besides being regarded as essentials for a broad spectrum of physiological functions during growth, differentiation and development, the literature of the last 10 years discloses that BRs have an anti-stress effect on plants at lower concentrations. These compounds have a wide range of biological activities that increase crop yields by changing plant metabolism and protecting plants from various types of stresses, like drought [22], salt [23], heat [24] and heavy metals [11-13, 25]. In the next topics, the ameliorative role of BRs in the growth of plants under heavy metal stresses will be discussed.

2.1 Effect of BRs on Seed Germination and Seedling Growth in the Presence of Excess Heavy Metal

Though there is some valuable information available on stress amelioration by BRs, the importance of this steroidal group of substances at the level of seed germination and seedling growth under metal stress is illusive.

Anuradha and Rao [26] examined the influence of BRs on seed germination and seedling growth in *Raphanus sativus* under cadmium toxicity. They reported that BRs supplementation ameliorated the toxic effect of the heavy metal and increased the percentage of seed germination and seedling growth (Table **1**).

Out of two analogues of BRs, 28-homobrassinolide (HBL) was found to be more effective than 24-Epibrassinolide (EBL) in stress alleviation. HBL (3 μM) alleviated the toxic effect of the heavy metal and increased the percentage of seed germination by 57%, over that of Cd and 20% over unstressed control. Similarly, supplementation of HBL (3 μM) caused an increase of 156%, 78% and 91% in shoot length (Table **1**), fresh weight and dry weight of seedlings, respectively, over those treated with Cd (Fig. **1**). BRs treatment also reduced the activities of peroxidase (POD) and ascorbic acid oxidase in heavy metal stressed seedlings. However, the stress amelioration effect of BRs was associated with enhanced level of proline (Fig. **1**).

Table 1. Effect of BRs on the percentage of seed germination (%) and seedling length (cm) of radish under cadmium stresses

Treatment	Seed germination (%)		Seedling length (cm)
	12 h	24 h	
Control	40.8 ± 4.05b	72.0 ± 2.76a	6.10 ± 0.28a
Cd 1.0mM	21.6 ± 2.61d	43.2 ± 2.90e	2.64 ± 0.39c
Cd 1.0mM + 1.0 mM EBL	25.6 ± 3.85d	59.2 ± 3.23c	3.72 ± 0.35c
Cd 1.0mM + 2.0 mM EBL	34.0 ± 4.65c	65.2 ± 2.38b	6.00 ± 0.37a
Cd 1.0mM + 3.0 mM EBL	48.8 ± 6.95a	66.4 ± 3.85b	5.76 ± 0.77b
Cd 1.0mM + 1.0 mM HBL	29.6 ± 3.58cd	49.2 ± 2.86d	5.80 ± 0.39b
Cd 1.0mM + 2.0 mM HBL	32.6 ± 3.86c	57.6 ± 2.71c	6.08± 0.26a
Cd 1.0mM + 3.0 mM HBL	50.4 ± 6.06a	65.6 ± 3.85b	6.76 ± 0.51a

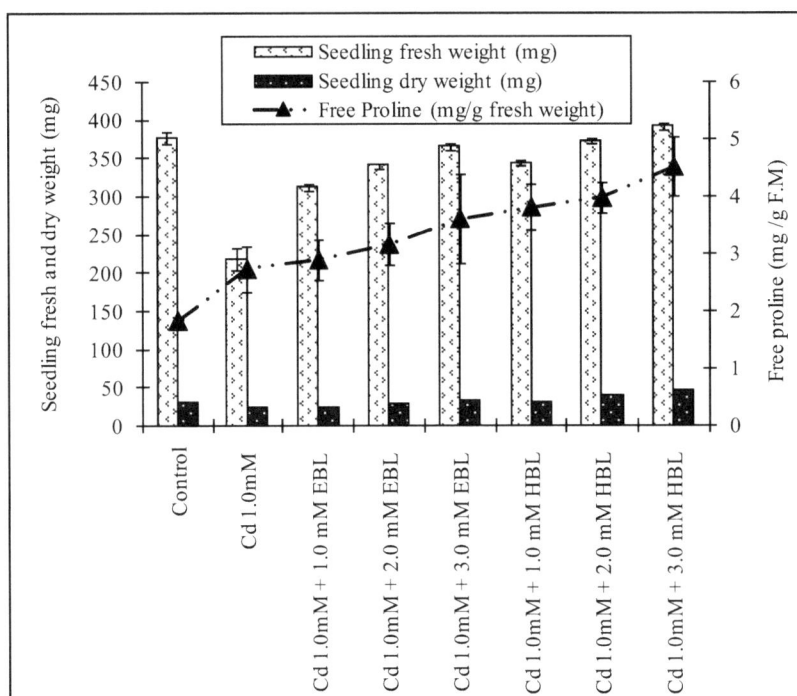

Figure 1: Effect of BRs on the seedling fresh and dry weight, and free proline content of radish, under cadmium stresses.

2.2 Effect of BRs on Growth Characteristics, Under Heavy Metal Stress

BRs were originally discovered as growth promoting substances isolated from pollen, and their role as plant hormone was confirmed in studies of photo morphogenesis. There is enough literature on the anti-stress role of BRs under heavy metal stress in terms of growth and development. This section is an attempt to compile the existing information on the anti-stress effect of BRs in plants grown under heavy metal stress.

Ali *et al.* [27] carried out a work on mung bean aiming at investigating a potential ameliorative role of BRs in plants subjected to aluminium (Al) stress. 1 week-old mung bean seedlings were subjected to aluminium (0.0, 1.0 or 10.0 mM) stress and then sprayed with 0 and 10^{-8} M 24-Epibrassinolide (EBL) or 28-homobrassinolide (HBL), after 14 days. Analysis of the plants 1 week after the BRs treatment revealed that the presence of aluminium in the nutrient medium caused a sharp reduction in the growth characteristics (length, fresh and dry mass of root and shoot), along with a reduction in relative leaf water content. The spray of EBL or HBL, in the absence of aluminium, strongly increased the above mentioned growth parameters, although BRs presented a weaker growth-stimulating effect in plants grown under aluminium stress (Fig. **2**).

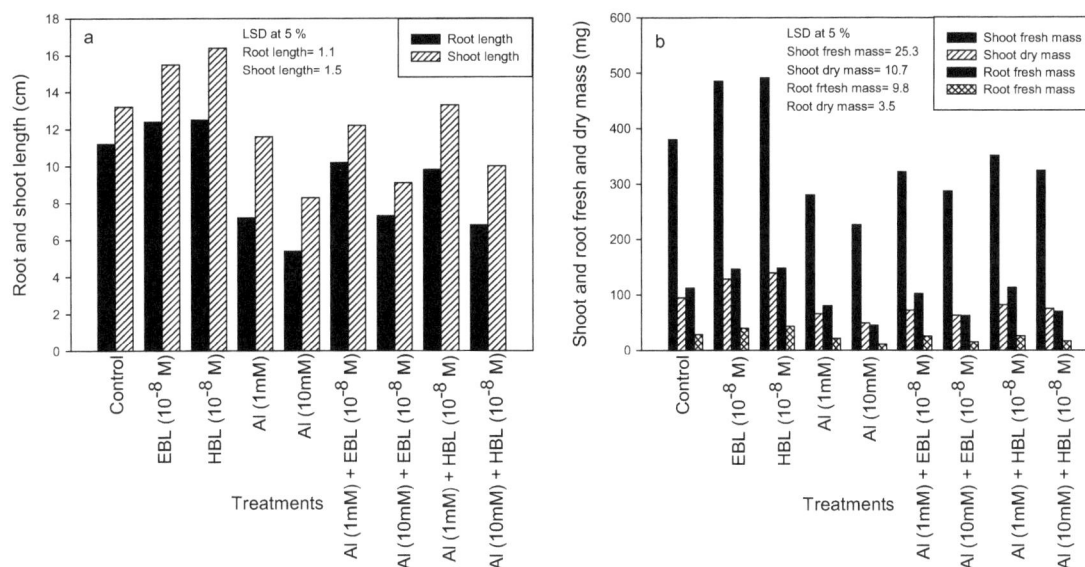

Figure 2: Effect of 24-epibrassinolide (EBL) and 28-homobrassinolide (HBL) on **(a)** root and shoot length (cm) and **(b)** fresh and dry mass of roots and shoots (mg) of aluminium (Al)-stressed mung bean seedlings.

Brassinosteroids generate such a response because of their involvement in the modification and/or manipulation of plasma membrane permeability under stress conditions [28]. Zhang *et al.* [39, 42] reported that two BRs, 28-homobrassinolide (HBL) and its direct precursor 28-homocastasterone (HCS), promote cell expansion of *Arabidopsis thaliana* suspension cells. They also showed that cell expansions induced by HBL and HCS are correlated with the amplitude of the plasma membrane hyperpolarization they elicited. HBL, which promoted the larger cell expansion, also provoked the larger hyperpolarization. We observed that membrane hyperpolarization and cell expansion were partially inhibited by the proton pump inhibitor erythrosin B, suggesting that proton pumps were not the only ion transport system modulated by the two BRs. Interestingly, while anion currents were inhibited by both HBL and HCS, outward rectifying K+ currents were increased by HBL but inhibited by HCS. In another study [33], the degree of resistance generated by BRs against metal stress in two tomato cultivars, at two different growth stages, was conducted using a shot gun approach. Seeds of K-25 and Sarvodya tomato cultivars were soaked in 100 μM CdCl$_2$ for 8 h (shotgun approach). Plants raised from CdCl$_2$-treated seeds showed significant reduction in leaf area, regardless the variety (Table **2**). However, the degree of toxicity generated by the Cd was more pronounced at 60 days after sowing in comparison to 90 days after sowing. The Cd treatment decreased leaf area by 40-fold and 20-fold, respectively for the Sarovdya and K-25 cultivars, compared to their corresponding controls. In addition to this, treatment with BRs (HBL/EBL) spray partially reversed the negative effect of Cd on the leaf area more effectively in K25 than in sarovdaya (Table **2**).

2.3 Effect of BRs on Yield and Dry Matter Production Under Heavy Metal Stress

Seeds of chickpea (*Cicer arietinum* (L.) were treated with 0, 50, 100 or 150 μM of cadmium in the form of cadmium chloride and sprayed with 0.01 μM of 28 homobrassinolide (HBL) 30 days after sowing. All of the yield characteristics (number of pods, seed yield, weight of 100 seeds, and seed protein content), except for the number of seeds per pod, were significantly affected by the treatment (Fig. **3**). The application of cadmium decreased the yield characteristics values and the loss was proportional to the increase in the concentration cadmium. Moreover, HBL significantly improved these characteristics and also neutralized the damaging effect of the metal. The hormone treatment to the plants already treated with 50 μM of Cd increased the values to a level that of the control. The plants treated with cadmium produced seeds with a lower level of protein, which decreased further in proportion to the concentration of the metal (Fig. **3**). However, the seeds developed on the HBL sprayed plant possessed significantly higher level of the protein. In addition to this, percent protein content in the seeds borned on the plants subjected to cadmium treatment (50 or 100 μM) followed with HBL were statistically equal to that of the control [12].

Table 2. Effect of BRs [28-homobrassinolide (HBL; 10^{-8} M) and 24-epibrassinolide (EBL; 10^{-8}M)] on the cadmium (100 μM) induced changes in leaf area (cm^2) in two different varieties of tomato (*Lycopersicon esculentum* Mill.) at 60 (24 hr after spray) and 90 days after sowing (DAS).

| | Leaf area (cm^2) | | | |
| | 60 DAS | | 90 DAS | |
	K-25	Sarvodya	K-25	Sarvodya
Control	7.51	6.72	8.81	7.68
HBL	7.83	6.91	11.4	10.0
EBL	7.94	7.06	12.7	10.7
$CdCl_2$	6.00	3.60	7.45	4.45
$CdCl_2$+HBL	6.21	3.77	9.17	5.10
$CdCl_2$+EBL	6.40	3.90	10.0	5.69
Mean	6.98	5.32	9.92	7.27

LSD at 5 %   Varieties (V) = 0.54   Varieties (V) = 0.49
             Treatments (T) = 1.01   Varieties (V) = 0.86
             V× T = 1.41   V× T = 1.21

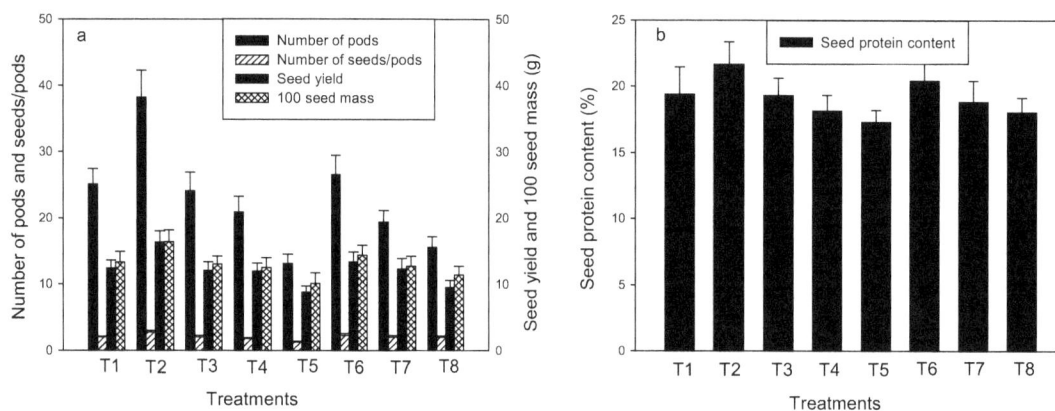

Figure 3: Effect of 0.01 μM 28-homobrassinolide (HBL) on cadmium (50, 100, 150 μM) induced changes in the **(a)** number of pods per plant, number of seeds per pod, seed yield (g) per plant, 100 seed mass (g) and **(b)** seed protein content (%) in *Cicer arietinum* at harvest [T1- Control; T2- HBL (0.01 μM); T3- Cd 50 μM; T4- Cd 100 μM; T5- Cd 150 μM; T6- Cd 50 μM + HBL (0.01 μM); T7- Cd 100 μM + HBL (0.01 μM); T8- Cd 150 μM + HBL (0.01 μM)].

A decrease in the seed yield per plant, the weight of 100 seeds and seed protein content in plants treated with cadmium is possibly a result of the poor growth and nodulation [12]. The effect of cadmium, however, can be overcome completely by HBL for plants fed with 50 μM of the metal. This counteract effect might be related to improved growth and photosynthetic characteristics [12], mineral and water uptake [30], assimilation of nitrate [31], photosynthesis [32] and protein synthesis [25].

Hayat *et al.* [33] studied the degree of resistance generated by BRs against metal stress in two tomato cultivars. Seeds (cv. K-25 and Sarvodya) were soaked in 100 μM $CdCl_2$ for 8 hrs. The resulting 59 day old seedlings were sprayed with 0.01 μM of 28 HBL or EBL. The treated plants were allowed to grow up to maturity (180 days after sowing), when yield characteristics of ripe fruits were recorded. Plants raised from seeds soaked in Cd before sowing showed significant reduction in all of the recorded yield characteristics (lycopene and β-carotene content, number of fruits and fruit yield per plant (Fig. **4**)), except for ascorbic acid content. However, BRs (HBL or EBL) spray significantly improved these characteristics and also neutralized the toxic effect of the metal tough more effectively in cultivar K-25 than in Sarvodya.

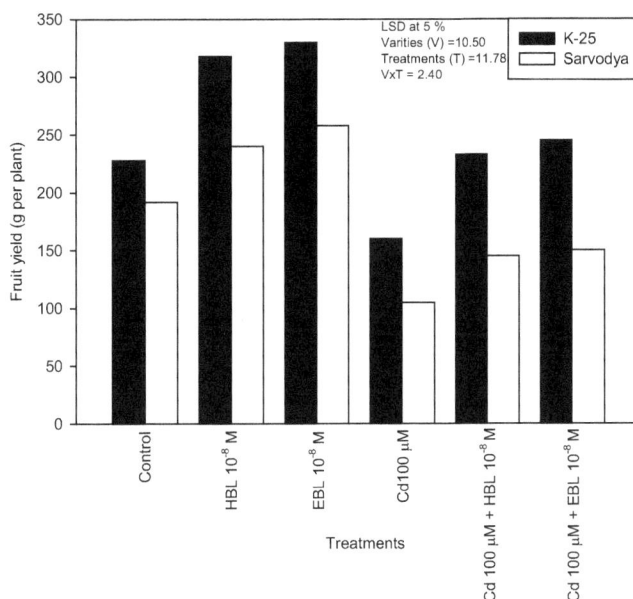

Figure 4: Effect of BRs [28-homobrassinolide (HBL; 10^{-8} M) and 24-epibrassinolide (EBL; 10^{-8}M)] on the cadmium (100 µM)-induced changes in fruit yield (g plants^{-1}) in two cultivars of tomato (*Lycopersicon esculentum* Mill.) at harvest.

2.4 Effect of BRs on Nitrogen Metabolism Under Heavy Metal Stress

Only a few studies have reported on the effect of BRs on the different aspects of nitrogen metabolism in leguminous crops under heavy metal stress.

Hasan *et al.* [12] carried out an experiment to explore the effect of exogenous BRs on nodule number per plant, leghaemoglobin content, nodule nitrogen content, and total carbohydrate content in nodules of *Cicer arietinum* submitted to cadmium stress. They found that application of cadmium resulted in significant decrease in the number of nodules per plant, which was proportional to the concentration of the metal (Fig. **5**). A significant increase in nodulation was noted in plants supplemented with HBL (0.01 µM). And, the negative effect of the lower concentration of the metal (50 µM) could be fully overcome, if the metal treatment was followed by HBL application (Fig. **5**).

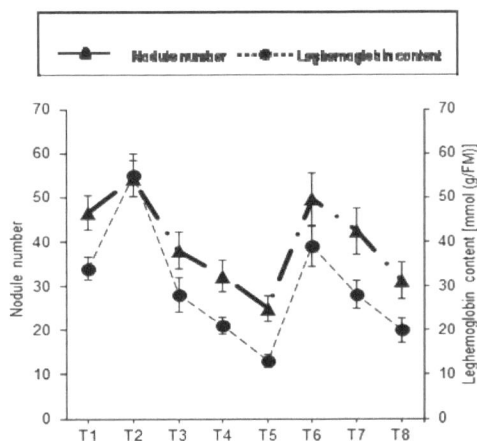

Figure 5: Effect of 28-homobrassinolide (HBL) on cadmium (50, 100, 150 µM) induced changes on nodule number and leghemoglobin content [mmol (g FM)$^{-1}$] in *Cicer arietinum* at 60 days, after sowing. [T1- Control; T2- HBL (0.01µM); T3- Cd 50µM; T4- Cd 100µM; T5- Cd 150µM; T6- Cd 50µM + HBL (0.01µM); T7- Cd 100µM + HBL (0.01µM); T8- Cd 150µM + HBL (0.01µM)].

Nodules grown on plants treated with HBL, alone, presented higher levels of nitrogen, carbohydrate (Fig. **6**) and leghaemoglobin, compared with control plants. However, nodules grown in plants treated with 100 and 150 µM cadmium presented reduced levels of nitrogen, carbohydrate and leghaemoglobin, while HBL treatment partially overcome the Cd-induced reduction in nitrogen, carbohydrate and leghemoglobin content in plants treated with 100 and 150 µM Cd.

2.5 Effect of BRs on Nitrate Reductase (NR) Acitivity Under Heavy Metal Stress

The process of reduction of nitrate is initiated by the enzyme nitrate reductase (NR). The level of NR activity decreased in plants of *Brassica juncea* treated with increased doses of Cu, Ni and Cd. However, the application of BRs (seed soaking/foliar spray) to heavy metal (Cu, Ni and Cd)-stressed plants attenuated the deleterious effects of heavy metals [11, 13, 34]. Treatment of *Cicer arietineum* with cadmium significantly inhibited NR activity, and the degree of inhibition was related to the cadmium concentration [12]. The deleterious effect of Cd was partially or completely counterbalanced by the subsequent treatment with HBL, depending on the concentration of the metal used in the treatments. A possible reason behind the ameliorative role of BR on NR activity of heavy metal-stressed plants could be an expression of the impact of BRs on translation or transcription [18] as it is well known that protein profile under above mentioned two conditions changes. However, a comparative proteomic study of resistant and sensitive varieties generally reflects the molecular mechanism of tolerance underlie. Another possibility is that HBL treatment might increase the endogenous level of NO_3 by acting at the membrane activating kinases and phosphatases thereby influencing the activity level of enzymes and also the cytosolic level of these genes induced by BRs applications [38].

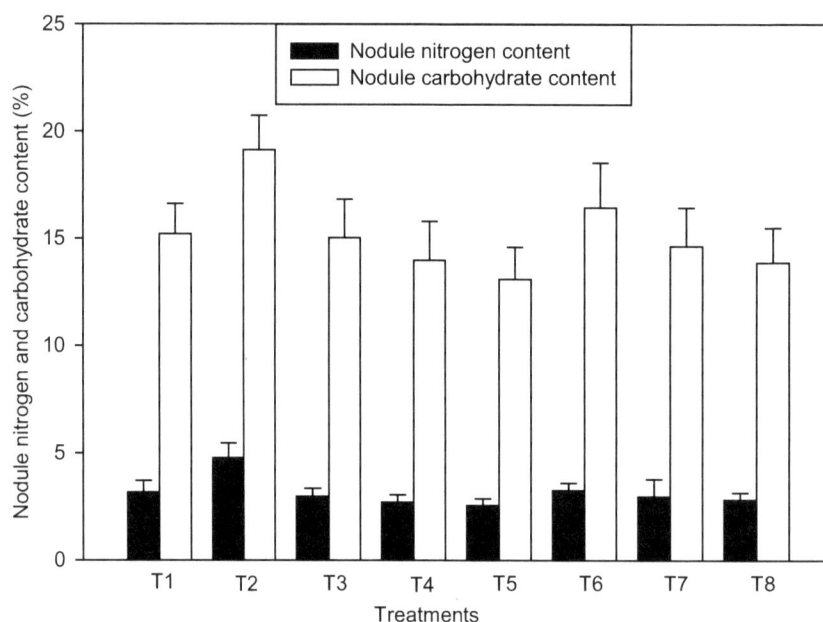

Figure 6: Effect of 28-homobrassinolide (HBL) on cadmium (50, 100, 150 µM) induced changes on nodule nitrogen and carbohydrate content (%) in *Cicer arietinum* at 60 days, after sowing. [T1- Control; T2- HBL (0.01µM); T3- Cd 50µM; T4- Cd 100µM; T5- Cd 150µM; T6- Cd 50µM + HBL (0.01µM); T7- Cd 100µM + HBL (0.01µM); T8- Cd 150µM + HBL (0.01µM)].

2.6 Effect of BRs on Chlorophyll Under Heavy Metal Stress

In a recent study [33], sensitive (Sarvodya) and tolerant (K-25) varieties of tomato cultivars at two different stages of growth under cadmium stress were shown to present larger reduction in chlorophyll content at the early stage of growth (*i.e.* 60 days after sowing (DAS)), when compared to plants at a later growth stage (90 DAS) (Table **3**). In contrast, the effect of cadmium was more efficiently reversed by foliar spray of HBL or EBL at 60 DAS, when compared to 90 DAS.

Table 3. Effect of BRs [28-homobrassinolide (HBL; 10^{-8} M) and 24-epibrassinolide (EBL; 10^{-8}M)] on the cadmium (100 μM) induced changes in SPAD value of chlorophyll in two different varieties of tomato (*Lycopersicon esculentum* Mill.) at 60 (24 hr after spray) and 90 days after sowing (DAS).

	SPAD Chlorophyll					
	60 DAS			90 DAS		
	K-25	Sarvodya	Mean	K-25	Sarvodya	Mean
Control	51.7	34.9	43.3	68.8	45.0	56.9
HBL	75.5	50.3	62.9	88.8	56.6	72.7
EBL	80.2	56.8	68.5	90.5	61.9	76.2
$CdCl_2$	45.1	25.5	35.3	62.3	35.1	48.7
$CdCl_2$+HBL	61.8	32.8	47.3	77.0	40.9	58.9
$CdCl_2$+EBL	66.4	35.7	51.0	79.6	43.5	61.5
Mean	63.4	39.3		77.8	47.1	

LSD at 5 %   Varieties (V) = 2.91   Varieties (V) = 2.59

Treatments (T) = 5.05   Varieties (V) = 4.48

V× T = 2.51   V× T = 2.90

Plants of *Vigna radiata* exposed to different levels of aluminum had lower chlorophyll content when compared to non-exposed plants, and the reduction in chlorophyll content was inversely related to increased doses of the metal. The BRs (HBL/EBL) treatment of unstressed plants resulted in strongly increased chlorophyll content, and, EBL more effectively enhanced chlorophyll content, when compared to HBL. In addition, BRs also reversed the effect of 1 mM aluminium stress (Fig.7).

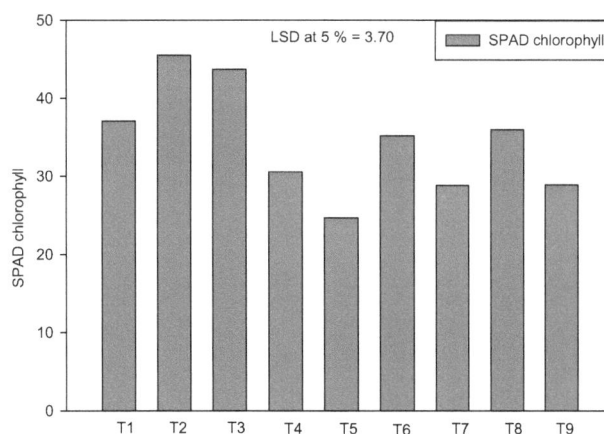

Figure 7: Effect of 24-epibrassinolide (EBL) and 28-homobrassinolide (HBL) on SPAD chlorophyll values in the leaves of aluminium (Al)-stressed mung bean seedlings. [T1- Control; T2- EBL (10^{-8}M); T3- HBL (10^{-8}M); T4- Al (1mM); T5- Al (10mM); T6- Al (1mM) + EBL (10^{-8}M); T7- Al (10mM) + EBL (10^{-8}M); T8- Al (1mM) + HBL (10^{-8}M); T9- Al (10mM) + HBL (10^{-8}M)].

2.7 Effect of BRs on Photosynthesis Under Excess Heavy Metals

Photosynthesis has also been found to be one of the most insightful processes in plants treated with different heavy metals. The metabolism of plants directly or indirectly depends on this process; any change in photosynthetic rate will automatically affect the metabolism of plants. Literatures revealed that BRs play a key role in enhancing the photosynthetic activity and their related attributes under different heavy metal stresses. Alam *et al.* [34] noted the effect of foliar application of HBL on nickel toxicity in *Brassica juncea*. They observed that the application of nickel inhibited the net photosynthetic rate which was more effective at their higher concentration. However, HBL applied alone enhanced the net photosynthetic rate in non-stressed plants, and, it partially neutralized the negative effect of Ni on photosynthesis (Fig. **8**).

In another study [13], different concentrations of HBL were used to analyze the net photosynthetic rate under different levels of copper in *Brassica juncea* plants. Among the three concentrations of HBL used, the lowest concentration (10^{-6}M) proved effective to ameliorate the ill-effect generated by excess copper. The reduction on the net photosynthetic rate, at the lowest dose of copper, was completely reversed by a pre-sowing seed soaking treatment with HBL (10^{-6}M), while HBL partially reversed the effect of the higher doses of the metal. Two analogues of BRs were sprayed on aluminum-stressed seedlings of mung bean to investigate the potential ameliorative effect of BR on aluminum-stressed seedlings. The metal stress decreased the photosynthetic rate and this inhibitory effect was proportional to the metal concentration. However, the BR treatment of unstressed plants strongly increased the net photosynthetic rate being EBL more effective than HBL. In addition, both BR analogues significantly enhanced net photosynthetic rate of plants grown under Al-stress (Fig. **9**).

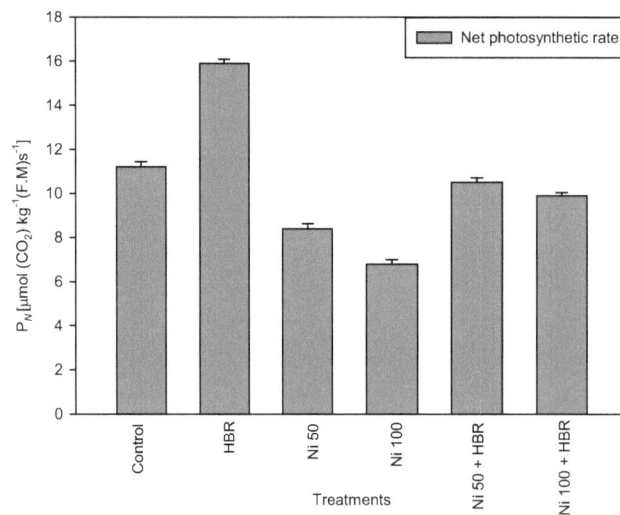

Figure 8: Effect of HBR (10^{-8} M) on nickel (Ni50 or Ni100 µM) induced changes in net photosynthetic rate (P_N) in *Brassica juncea* cv. T-59 at 40 d after sowing.

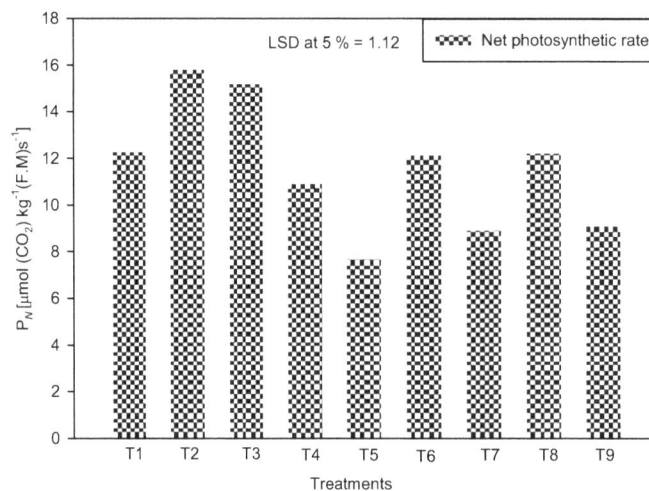

Figure 9: Effect of 24-epibrassinolide (EBL) and 28-homobrassinolide (HBL) on the photosynthetic rate (µmol CO_2 Kg^{-2} s^{-1}) in the leaves of aluminium (Al)-stressed mung bean seedlings. [T1- Control; T2- EBL (10^{-8}M); T3- HBL (10^{-8}M); T4- Al (1mM); T5- Al (10mM); T6- Al (1mM) + EBL (10^{-8}M); T7- Al (10mM) + EBL (10^{-8}M); T8- Al (1mM) + HBL (10^{-8}M); T9- Al (10mM) + HBL (10^{-8}M)].

2.8 Effect of BRs on Antioxidant System Under Heavy Metal Stress

There is ample evidence that exposure of plants to different heavy metals results in overproduction of reactive oxygen species (ROS). To combat the negative consequences of ROS under heavy metal stress, plants develop the ability to increase their antioxidant protection. There have been attempts to increase the level of the antioxidant system under different environmental stresses, including heavy metals. There are many reports which showed that application of BRs resulted in the increased activity of the antioxidant enzymes viz. CAT (catalase), POX (peroxidase), SOD (superoxide dismutase) under different heavy metal stresses. For example, BRs enhanced the level of antioxidant system under cadmium stress in *Brassica juncea* [11]. Hasan *et al.* [12] established the relationship between antioxidant system (CAT, POX, and SOD) and BRs in the presence of excess cadmium in chickpea. They observed a significant increase in the activity of antioxidative enzymes (catalase, peroxidase, superoxide dismutase) in response to cadmium and/or HBL (Fig. **10**). This increase in activity was directly related to increased levels of the metal (50-150 μM). Moreover, the foliar spray of HBL to the cadmium-treated plants had an additive effect on the enzyme's activity. It was concluded by Hasan *et al.* [12] that this elevated antioxidative defense system made the stressed plants resistant to cadmium stress.

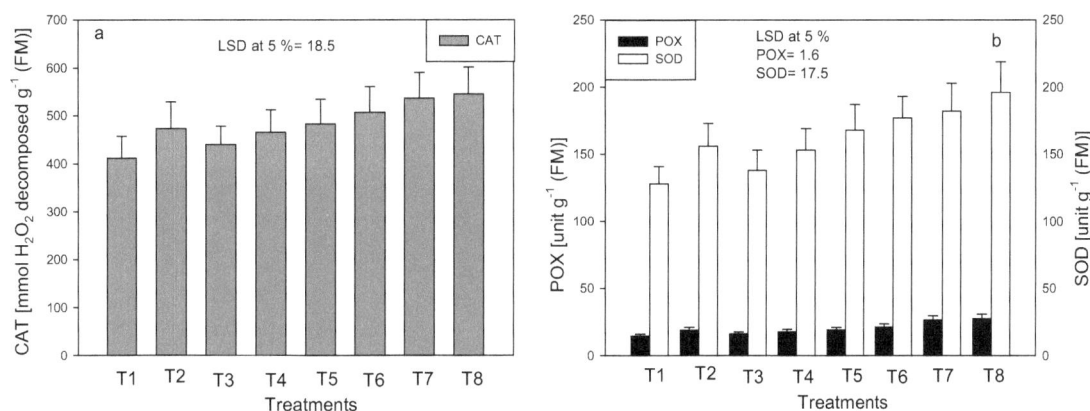

Figure 10: Effect of 28-homobrassinolide (HBL) on cadmium (50, 100, 150 μM) induced changes in the (a)activity of catalase [mmol H_2O_2 decomposed g^{-1} (FM)], (b) peroxidase [unit g^{-1} (FM)] and superoxide dismutase (SOD) [units g-1 fresh mass] in *Cicer arietinum* at 60 days, after sowing. [T1- Control; T2- HBL (0.01μM); T3- Cd 50μM; T4- Cd 100μM; T5- Cd 150μM; T6- Cd 50μM + HBL (0.01μM); T7- Cd 100μM + HBL (0.01μM); T8- Cd 150μM + HBL (0.01μM)]

Ali *et al.* [27] investigated the possible relationship between the antioxidant system and the degree of resistance induced by BRs against the stress generated by aluminum in mungbean seedlings. Similarly to what had been observed for chickpea, the activity of antioxidative enzymes (CAT, POX and SOD) in leaves increased in plants subjected to the Al-stress. Moreoever, the subsequent treatment of the seedlings with BRs (HBL or EBL) further enhanced the activity of the antioxidative enzymes. Among the two analogues of BRs, HBL proved to be more effective, than EBL.

Hayat *et al.* [33] observed that heavy metal sensitive (Sarvodya) and tolerant (K-25) varieties of tomato, at two different stages of growth (*i.e.* 60 and 90 days after sowing), showed significantly different responses towards antioxidant system, to BRs spray treatments. Activity of antioxidative enzyme SOD increased in response to both the metal and the hormone treatments, being the K-25 cultivar more responsive to the treatments compared to the Sarvodya cultivar (Table **4**).

3. RESPONSE OF BRs IN HEAVY METAL DETOXIFICATION

It is now well documented that the excess of heavy metal toxicity can induce a variety of adaptive responses in plants. The chelation of the heavy metal ion by a ligand is an ever-present mechanism for heavy metal detoxification. There are different such ligands that include organic acids, amino acids, peptides and polypeptides. Peptide ligands include the phytochelatins (PC), small gene-encoded cysteine-rich polypeptides. These heavy metal-binding peptides derive from glutathione and have the general structure (g-Glu-Cys)n-Gly, where n has been reported as being as high as 11, but it is generally in the 2–5 range. PC detoxifies intracellular metals by binding them through thiolate coordination [35]. BRs have been shown to stimulate the synthesis of PC in *Chlorella vulgaris* cells treated with lead. This stimulatory effect of BRs on PC synthesis was arranged in the following order: brassinolide (BL) > 24-epiBL > 28-homo-BL > castasterone (CS) > 24-epiCS > 28-homoCS [36].

Table 4. Effect of BRs [28-homobrassinolide (HBL; 10^{-8} M) and 24-epibrassinolide (EBL; 10^{-8}M)] on the cadmium (100 μM)-induced changes in superoxide dismutase (SOD) activity [units g^{-1} (F.M.)] in two different varieties of tomato (*Lycopersicon esculentum* Mill.) at 60 (24 hr after spray) and 90 days after sowing (DAS).

	Superoxide dismutase activity [units g^{-1} (F.M.)]					
	60 DAS			90 DAS		
	K-25	Sarvodya	Mean	K-25	Sarvodya	Mean
Control	128	110	119	147	121	134
HBL	158	135	146	169	135	152
EBL	163	143	153	174	140	157
CdCl$_2$	155	126	140	164	133	148
CdCl$_2$+HBL	184	149	166	201	153	177
CdCl$_2$+EBL	198	160	179	213	162	187
Mean	164	137		178	140	

LSD at 5 %   Varieties (V) = 0.30   Varieties (V) = 16.30
                                     Treatments (T) = 0.94   Varieties (V) = 28.23
                                     V× T = 0.99   V× T = 0.39

The cultures of *Chlorella vulgaris* treated with BRs and heavy metals showed lower accumulation of heavy metals than cultures treated with metals alone. The inhibitory effect of BRs mixed with different heavy metals on the metal accumulation was arranged in the following order: zinc > cadmium > lead > copper. The BRs-induced inhibition of metal accumulation resulted in increased growth and development of *Chlorella vulgaris*. Application of BRs to heavy metals stressed *Chlorella vulgaris* cultures not only reduced the impact of heavy metals stress on growth, but it also prevented chlorophyll, sugar and protein loss and increased PC synthesis. An inverse relation has been observed between the toxicity based on level of metal applied and concentration of BRs to stimulate the growth of plant cells. However, reaching its optimal activity for BRs curve exponentially stabilizes.

Biosorption is another mechanism for metal removal. This mechanism is generally based on physico-chemical interactions between metal ions and the functional groups present on the cell surface, such as electrostatic interactions, ion exchange and metal ion chelation or complexation and functional groups most commonly implicated in such interactions include carboxylate, hydroxyl, amine and phosphoryl groups present within cell wall components such as polysaccharides, lipids and proteins. In addition, biosorption can be affected by pH and the presence of other ions in the medium which may precipitate heavy metals as insoluble salts, but it is unaffected by metabolic inhibitors or light/dark cycles [37]. Water pH is an important factor directly affecting the toxicity of metals in algae such as unicellular *Chlorella* sp. It is known that heavy metal toxicity increases with decreasing pH. Then, acidity or alkalinity of the medium can modulate the toxicity of heavy metals. Lower pH may increase the bioavailability of metal ions resulting in increased toxicity.

Growth in *Chlorella vulgaris* as suggested by Bajguz [25], stimulated by BRs, depends at least partly upon apoplastic low pH or acid induced loosening of cell wall. Brassinosteroids regulate plasma membrane anion channels in addition to proton pumps during expansion of *Arabidopsis thaliana* suspension cells [39, 42]. The effects of BRs on proton secretion are associated with an early hyperpolarization of the transmembrane electrical potential. BR-induced proton excretion may also be stimulated by the presence of Pb in the medium [25].

REFERENCES

[1]   Steffens GL. U.S. Department of Agriculture Brassins Project: 1970–1980. In: Cutler HG, Yokota T, and Adam G, Eds. Brassinosteroids - Chemistry, Bioactivity and Applications. Washington: American Chemical Society 1991; pp. 2-17.

[2]   Mitchell JW, Mandava NB, Worley JF, Plimmer JR, Smith MV. Brassins - A new family of plant hormones from rape pollen. Nature 1970; 225: 1065–1066.

[3]   Grove MD, Spencer FG, Rohwededer WK, *et al.* Brassinolide, a plant growth promoting steroid isolated from *Brassica napus* pollen. Nature 1979; 281: 216–217.

[4]   Bajguz A, Tretyn A. The chemical characteristic and distribution of brassinosteroids in plants. Phytochem 2003; 62: 1027-1046.

[5]   Clouse SD, Sasse JM. Brassinosteroids: Essential regulators of plant growth and development. Annu Rev Plant Physiol Plant Mol Biol 1998; 49: 427-451.

[6]   Ohashi-Ito K, Fukuda H. HD-zip III homeobox genes that include a novel member, ZeHB-13 (*Zinnia*)/ATHB-15 (*Arabidopsis*), are involved in procambium and xylem cell differentiation. Plant Cell Physiol 2003; 44: 1350-1358.

[7]   Symons GM, Davies C, Shavrukov Y, *et al.* Grapes on steroids. Brassinosteroids are involved in grape berry ripening. Plant Physiol 2006; 140: 150-158.

[8]   Szekeres M, Nemeth K, Koncz-Kalman Z, *et al.* Brassinosteroids rescue the deficiency of CYP90, a cytochrome P450, controlling cell elongation and de-etiolation in *Arabidopsis*. Cell 1996; 85: 171-182.

[9]   Clouse SD. Molecular genetic studies confirm the role of brassinosteroids in plant growth and development. Plant J 1996; 10: 1-8.

[10]   Vardhini BV, Anuradha S, Rao SSR. Brassinosteroids - new class of plant hormones with potential to improve crop productivity. Indian J Plant Physiol 2006; 11: 1-12.

[11]   Hayat S, Ali B, Hasan SA, Ahmad A. Brassinosteroid enhanced the level of antioxidants under cadmium stress in *Brassica juncea*. Environ Exp Bot 2007; 60: 33–41.

[12]   Hasan SA, Hayat S, Ali B, Ahmad A. 28-homobrassinolide protects chickpea (*Cicer arietinum*) from cadmium toxicity by stimulating antioxidants. Environ Poll 2008; 151: 60–66.

[13]   Fariduddin Q, Yusuf M, Hayat S, Ahmad A. Effect of 28-homobrassinolide on antioxidant capacity and photosynthesis in *Brassica juncea* plants exposed to different levels of copper. Environ Exp Bot 2009; 66: 418-424.

[14]   Kinoshita T, Cano-Delgado A, Seto H, *et al.* Binding of brassinosteroids to the extracellular domain of plant receptor kinase BRI1. Nature 2005; 433: 167-171.

[15]   Sakurai A. Biosynthesis. In: Sakurai A, Yokota T, Clouse SD, Eds. Brassinosteroids: Steroidal Plant Hormones. Tokyo: Springer Verlag 1999; pp. 91-111.

[16]   Khripach VA, Zhabinskii VN, de Groot AE. Brassinosteroids. A New Class of Plant Hormones. San Diego: Academic Press 1999; 456p.

[17]   Li J, Chory J. Brassinosteroid actions in plants. J Exp Bot 1999; 332: 275–282.

[18]   Khripach V, Zhabinskii V, de Groot A. Twenty years of brassinosteroids: steroidal plant hormones warrant better crops for the XXI century. Ann Bot 2000; 86: 441–447.

[19]   Müssig C, Altmann T. Brassinosteroid signalling in plants. Trends Endocrin Metab 2001; 12: 398–402.

[20]   Zullo MAT, Adam G. Brassinosteroid phytohormones-structure, bioactivity and applications. Braz J Plant Physiol 2002; 14: 143–181.

[21]   Fujioka S, Yokota T. Biosynthesis and metabolism of brassinosteroids. Annu Rev Plant Physiol Plant Mol Biol 2003; 54: 137–164.

[22]   Upreti KK, Murti GSR. Effects of brassinosteroids on growth, nodulation, phytohormone content and nitrogenase activity in French bean under water stress. Biol Plant 2004; 48: 407–411.

[23] Ozdemir F, Bor M, Demiral T, Turkan I. Effects of 24-epibrassinolide on seed germination, seedling growth, lipid peroxidation, proline content and antioxidative system of rice (*Oryza sativa* L.) under salinity stress. Plant Growth Regul 2004; 42: 203–211.

[24] Dhaubhadel S, Browning KS, Gallie DR, Krishna P. Brassinosteroid functions to protect the translational machinery and heat shock protein synthesis following thermal stress. Plant J 2002; 29: 681–691.

[25] Bajguz. Blockade of heavy metals accumulation in *Chlorella vulgaris* cells by 24-epibrassinolide. Plant Physiol Biochem 2000; 38: 797–801.

[26] Anuradha S, Rao SSR. The effect of brassinosteroids on radish (*Raphnus sativus* L.) seedlings growing under cadmium stress. Plant Soil Environ 2007; 53: 465-472.

[27] Ali B, Hasan SA, Hayat S, *et al.* A role for brassinosteroids in the amelioration of aluminum stress through antioxidant system in mung bean (*Vigna radiata* L. Wilczek). Environ Exp Bot 2008; 62: 153-159.

[28] Hamada K. Brassinolide: some effects of crop cultivations. Conf. Proc. Int. Seminar Tokyo, Japan. Plant Growth Regul 1986; 15: 65–69.

[29] Hayat S, Bajguz A. Effect of brassinosteroids on the plant responses to environmental stresses. Plant Physiol Biochem 2009; 47: 1-8.

[30] Ali B, Hayat S, Ahmad A. Response of germinating seeds of *Cicer arietinum* to 28-homobrassinolide and/or potassium. Gen Appl Plant Physiol 2005; 31: 55–63.

[31] Mai Y, Lin S, Zeng X, Ran R. Effect of brassinolide on nitrate reductase activity in rice seedlings. Plant Physiol Commun 1989; 2: 50-52.

[32] Yu JQ, Huang LF, Hu WH, *et al.* A role for brassinosteroids in the regulation of photosynthesis in *Cucumis sativus*. J Exp Bot 2004; 55: 1135–1143.

[33] Hayat S, Hasan SA, Hayat Q, Ahmad A. Brassinosteroids protect *Lycopersicon esculentum* from cadmium toxicity applied as shotgun approach. Protoplasma 2010; 239: 3-14.

[34] Alam MM, Hayat S, Ali B, Ahmad A. Effect of 28-homobrassinolide on nickel induced changes in *Brassica juncea*. Photosynthetica 2007; 45: 139–142.

[35] Cobbett C, Goldsbrough P. Phytochelatins and metallothioneins: roles in heavy metal detoxification and homeostasis. Annu Rev Plant Physiol Plant Mol Biol 2002; 53: 159–182.

[36] Bajguz A. Brassinosteroids and lead as stimulators of phytochelatins synthesis in *Chlorella vulgaris*, J Plant Physiol 2002; 159: 321–324.

[37] Vilchez C, Garbayo I, Lobato MV, Vega JM. Microalgae-mediated chemicals production and wastes removal. Enz Microbiol Technol 1997; 20: 562–572.

[38] Taiz L, Zeiger E. Brassinosteroids. In: Plant Physiology. Sunderland, M.A., USA: Sinauer Associates Inc 2004; pp: 617-634.

[39] Zhang Z, Ramirez J, Reboutier D, *et al.* Brassinosteroids regulate plasma membrane anion channels in addition to proton pumps during expansion of *Arabidopsis thaliana* cells. Plant Cell Physiol 2005; 46: 1494–1504.

[40] Divi UK, Rahman T, Krishna P. Brassinosteroid-mediated stress tolerance in *Arabidopsis* shows interactions with abscisic acid, ethylene and salicylic acid pathways. BMC Plant Biol 2010; 10: 151-165.

[41] Arteca RN, Arteca JM. Effects of brassinosteroid, auxin, and cytokinin on ethylene production in *Arabidopsis thaliana* plants. J Exp Bot 2008; 59: 3019–3026.

[42] Hayashi Y, Nakamura S, Takemiya A, Takahashi Y, Shimazaki K, Kinoshita T. Biochemical characterization of *in vitro* phosphorylation and dephosphorylation of the plasma membrane H$^+$-ATPase. Plant Cell Physiol. 2010; 51: 1186–1196.

CHAPTER 6

Antiviral Properties of Brassinosteroids

Mónica B. Wachsman[*] and Viviana Castilla

Laboratorio de Virología. Departamento de Química Biológica. Facultad de Ciencias Exactas y Naturales, Universidad de Buenos Aires, Ciudad Universitaria, Pabellón 2, Piso 4, C1428EGA, Buenos Aires, Argentina

Abstract: In this chapter, we reviewed the antiviral activity of natural and synthetic brassinosteroids (BRs). Brassinolide and other natural BRs such as 28-homocastasterone present a broad antiviral spectrum against RNA and DNA viruses. Since the concentration of BRs in the plant tissue is very low, isolation of these compounds from plant sources is impractical, and they had to be obtained by chemical synthesis. The antiviral activity of a group of synthetic analogues against Herpes Simplex Type 1 and 2 (HSV-1 and HSV-2), Measles (MV), Vesicular Stomatitis Virus (VSV), Polio Virus (PV) and Arena Viruses was also determined. Several of the tested compounds showed selectivity indexes 10- to 18-fold higher than ribavirin, a broad spectrum antiviral compound, for Junín Virus (JUNV), a moderate activity against HSV-1 / HSV-2 and good anti-MV, anti-PV and anti-VSV activity, with antiviral selectivity Index (SI) values also higher than ribavirin, a reference drug. Structure activity relationship Studies were performed in order to determine which stereochemistry, type and position of functional groups are needed to develop a potent and selective class of viral inhibitors. The antiviral mode of action of the BRs against HSV, JUNV and VSV was also investigated. For all assayed viruses, the antiviral compounds adversely affect virus protein synthesis and mature viral particle formation.

Keywords: Herpes Simplex Virus, Junín virus, Measles virus, Poliovirus, Ribavirin, Vesicular Stomatitis virus.

INTRODUCTION

Compared to the antibiotic therapy available for bacterial infections, antiviral chemotherapy is still in its infancy. There are many viral diseases for which no effective drugs or vaccines exist. Unlike most bacteria, which can have an independent existence outside host cells, viruses are intracellular parasites which require host cell mechanisms in order to replicate. It has proved difficult to find compounds that can selectively block viral replication without interference to the normal cellular processes and thus with significant toxicity to the host. Many antiviral agents in clinical use today or undergoing active evaluation, known as inhibitors of pathogenic virus replication, are synthetic chemical analogues of nucleosides. Because of their close chemical similarity to naturally occurring analogues or substrates, they may be falsely incorporated in the viral biosynthetic process, thereby disrupting viral replication. Up today, the best antiviral compound available is acyclovir. It is highly selective against Herpes Simplex Virus (HSV) and to a lesser extent against other members of the Herpes virus family [1, 2]. It is clear that a detailed understanding of the molecular targets encoded by viruses can be of potential use in the development of specific antiviral therapy. One must take into account that a successful anti-viral drug should: (i) interfere with a virus-specific function (either because the function is unique to the virus or the similar host function is much less susceptible to the drug) or (ii) interfere with a cellular function so that the virus cannot replicate. To be specific, the antiviral drug must only kill virus-infected cells. This could be done by restricting drug activation to virus-infected cells. Obviously, a good drug must show much more toxicity to the virus than to the host cell [2].

Antiviral therapy is now available against several viral diseases like inhibitors of viral proteases or viral integrases designed to interrupt specifically Human Immunodeficiency Virus (HIV) replication [3] or in the case of Influenza virus to mimic the substrate for sialidase enzymatic activity [4, 5]. Unfortunately, viruses

---

*Address correspondence to Mónica B. Wachsman: Laboratorio de Virología. Departamento de Química Biológica. Facultad de Ciencias Exactas y Naturales, Universidad de Buenos Aires, Ciudad Universitaria, Pabellón 2, Piso 4, C1428EGA, Buenos Aires, Argentina; E-mail: wachsman@qb.fcen.uba.ar

respond to antiviral treatment with a rapid selection of drug resistant viral mutants. Therefore, more and better agents must certainly be available in the near future.

Plants have evolved constitutive and inducible defense mechanisms by producing a vast array of hormones and secondary metabolites against various microbial pathogens [6]. It is therefore probable that antiviral compounds would occur in plants as part of their innate defense and these compounds constitute a promissory group of molecules for screening of new antiviral agents.

PLANT STEROIDS

Sterols are membrane components found in all eukaryotic organisms which regulate fluidity and permeability of phospholipid bilayers.

Plant sterols have been extensively studied in past years with a major focus on biosynthetic and biochemical aspects [7, 8]. Minor proportions of these compounds serve as precursors to steroid derivatives, and were recognized as a new class of plant growth regulators called brassinosteroids (BRs).

Brassinolide was the first identified plant steroid with hormonal activities (Fig. 1). Since then, many other structurally related steroidal compounds with growth-promoting activities have been isolated from plants; Now, about 60 of these compounds are known [9] and they are collectively referred to as BRs. BRs contain the typical 5α-cholestan steroidal skeleton with fused A, B, C, and D rings and an alkyl side chain at C-17, not found in mammalian steroidal hormones.

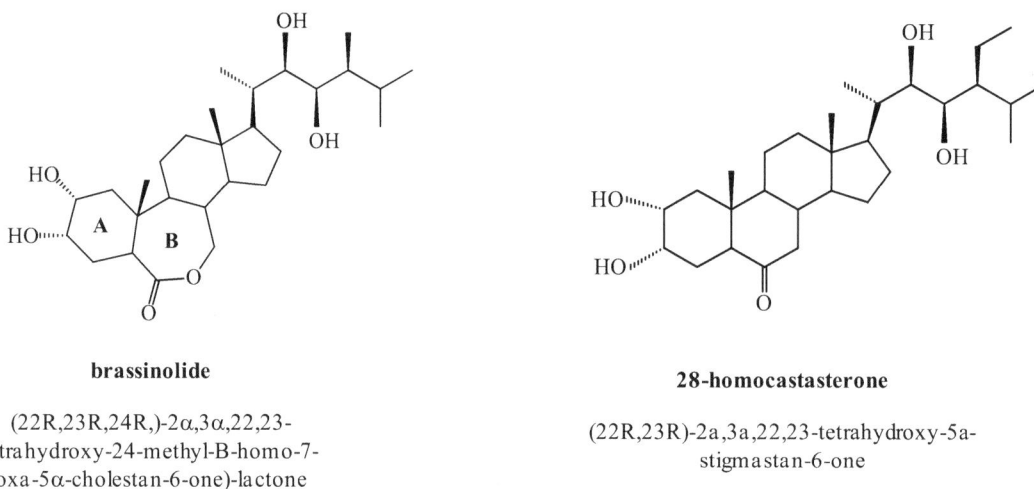

brassinolide

(22R,23R,24R,)-2α,3α,22,23-
tetrahydroxy-24-methyl-B-homo-7-
oxa-5α-cholestan-6-one)-lactone

28-homocastasterone

(22R,23R)-2a,3a,22,23-tetrahydroxy-5a-
stigmastan-6-one

Figure 1: Structural formulae of brassinolide and 28-homocastasterone.

Preliminary Antiviral Studies with Steroidal Molecules

At the start of our investigation, there were not previous reports on the biological effect of natural or synthetic analogs of BRs on cell cultures from mammalian origin, neither on its capability to interfere with the replication of animal viruses. Although a number of biologically active steroids bearing unusual side chains, isolated from marine sponges, had been studied for their antiviral activity. Orthoesterols A, B and C have been reported to be active Against Feline Leukaemia Virus (FeLV), Mouse Influenza Virus and Mouse Corona Virus [10]. Weinbersterols A and B, also isolated from a marine sponge (*Petrosia weinbergi*), exhibited *in vitro* activity against FeLV with an EC$_{50}$ (50% effective concentration or compound concentration that reduce by 50 % virus replication, with respect to an untreated control) values of 40 and 52 µg/mL, respectively. Weinbersterols A and B also showed activity against HIV with an EC$_{50}$ value of 10 µg/mL [11]. It has also been reported by Comin *et al.* [12] that sulfated steroids, isolated from the cold water ophiuroids, *Ophioplocus januarii*, and *Astrotoma agassizii,* show antiviral activity against human pathogenic viruses like HSV-2, JUNV and PV. When used in concentrations of 40 or 80 µg/mL, most of the assayed compounds induce a reduction in the percentage of virus plaque formation lower than 50 %, compared to untreated control cultures.

Arthan *et al.* [13] demonstrated that steroidal glycosides isolated from *Solanum torvum*, a small shrub distributed widely in Thailand, display antiviral activity against HSV-1. On the other hand, dehydroepiandrosterone (DHEA), an animal androgenic steroidal hormone, and its synthetic derivatives inhibit the multiplication of JUNV, VSV, adenovirus (AdV), HIV and Japanese encephalitis virus (JEV) [14-20].

ANTIVIRAL ACTIVITY OF NATURAL BRs

In the last years, our laboratory has described the inhibitory action of several natural and synthetic BRs against different animal viruses [21-26]. It was first demonstrated that two natural BRs, brassinolide and 28-homocastasterone **(1a)** (Fig. 1), display antiviral activity against poliovirus (PV), HSV-1 and HSV-2, measles virus (MV), vesicular stomatitis virus (VSV) and the arenaviruses: Junin (JUNV), Tacaribe (TCRV) and Pichinde (PICV) (Table 1) [25], with higher inhibition values than previously reported assays where other natural steroidal molecules were tested [10-13]. A series of 28-homoBRs analogues synthesized at the Laboratory of Organic Chemistry of the Faculty of Science of the University of Buenos Aires, allowed us to investigate the potential antiviral activity of this kind of steroids.

Table 1. Antiviral activity of natural BRs against several RNA and DNA viruses.

Viruses	Inhibition (%)	
	Brassinolide	**28-homocastaterone**
PV type I	96	85
VSV Indiana strain	100	23
JUNV IV $_{4454}$ strain	74	79
TCRV TRLV $_{11573}$ strain	55	99
PICV AN$_{3739}$ strain	67	98
HSV-1 F strain (tk$^+$)	96	50
HSV-1 B2006 strain (tk$^-$)	100	35
HSV-2 G strain (tk$^+$)	98	48
MV Brasil/001/91	100	50

[†] The assayed concentrations of BRs were not toxic to the host cells.

Vero cells were infected at a multiplicity of infection (moi, number of *virus* particles per cell) of 1 with the different viruses. After 1 h adsorption at 37°C, the cultures were covered with MM (maintenance medium) or with MM containing brassinolide at a concentration of 1 µM (the maximum available concentration) or 28-homocastasterone, at a concentration of 40 µM. After 18-24 h of incubation at 37°C, supernatants were harvested and titrated by a plaque assay.

Animal viruses tested for their susceptibility to BRs analogues comprised four RNA viral families, *Paramyxoviridae, Arenaviridae, Picornaviridae, Rhabdoviridae* and one DNA virus family *Herpesviridae*, all of them important human or animal pathogens. MV, a member of *Paramyxoviridae*, causes infection through an acute respiratory infection and immunosuppression. Despite the generalized use of an effective live attenuated vaccine, MV continues to contribute to high infant mortality in underdeveloped countries [27]. JUNV, TCRV, PICV are members of the *Arenaviridae* family. JUNV causes a severe disease in humans known as Argentine hemorrhagic fever [28]. PV, a member of the *Picornaviridae* family, still represents a relevant health problem in some countries of the world, despite a dramatic decrease in its incidence as a result of intensive vaccination programs in developed countries [29]. VSV, a member of the *Rhabdoviridae* family, produces a vesicular disease in foot-and-mouth disease-free countries in cattle, horses, swine and lambs, causing thousands of outbreaks every year [30]. HSV-1 and HSV-2 are serious human pathogens. HSV-1 is normally associated with orofacial infections and encephalitis, whereas HSV-2 usually causes genital infections and can be transmitted from infected mothers to neonates [31]. Our studies were performed with a total of 37 steroidal compounds, among them 27 were analogues with BR-like structure synthesized from stigmasterol (Fig. **2**).

Cytotoxic and Antiviral Activities of BRs Derivatives

After the promissory results obtained with brassinolide and 28-homocastasterone we decided to test a collection of synthetic BRs in order to test its potency as virus inhibitors. Structural formulae of all synthetic analogues used are shown in Table **2**. The syntheses of the compounds have been described elsewhere [25]. Afterwards, the synthesized BRs where tested for cytotoxic and antiviral activity.

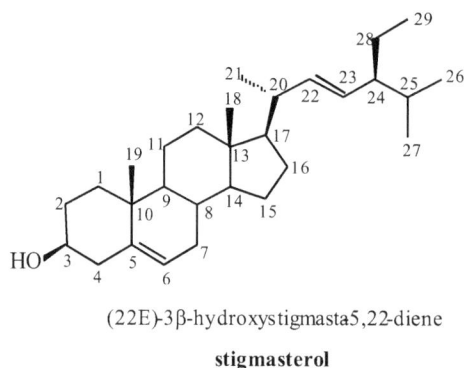

(22E)-3β-hydroxystigmasta5,22-diene

stigmasterol

Figure 2: Structural formula of stigmasterol.

Cytotoxicity Studies

Cytotoxicity of all compounds was determined on stationary Vero cell monolayers, using a standard method based on the calculation of 50% cell viability, after 24 h of incubation at 37°C, in the presence of different concentrations of each derivative [32]. Toxicity studies of synthetic BRs are shown in Table **2**. Stigmasterol was included in Table **2** because it is a natural sterol, and also the starting material for all of the synthetic derivatives. Table **2** also includes in the last rows the cytotoxicity values of acyclovir (ACV) and ribavirin (RBV), the standard compounds used as antiviral agents against HSV and MV, or JUNV, respectively.

Table 2. Cytotoxicity and antiviral activity of BRs.

Structural formula	Compound	IS**				
		CC_{50}* (µM)	HSV-1	HSV-2	MV	JV
	stigmasterol	479	2	2	4	8

Structural formula	Compound	IS**				
		CC$_{50}$* (µM)	HSV-1	HSV-2	MV	JV
	1a	63	2	ND	2	18
	1b	259	26	ND	4	61
	9a	427	28	14	57	28
	28a	1259	19	29	ND	ND
	14	158	2	2	4	27
	25	>848	10	10	ND	ND
	3a	270	11	ND	7	39
	3b	226	10	ND	8	232

Structural formula	Compound	IS**				
		CC$_{50}$* (μM)	HSV-1	HSV-2	MV	JV
	5a	114	4	ND	2	310
	5b	152	6	ND	4	72
	17a	215	4	2	32	4
	17b	43	8	2	33	4
	30a	1621	39	39	ND	ND
	26a	<108	<4	<5	ND	ND
	22a	301	16	16	8	16
	2a	263	<1	ND	2	200

Structural formula	Compound	IS**				
		CC$_{50}$* (μM)	HSV-1	HSV-2	MV	JV
	2b	502	16	ND	6	228
	4a	230	17	ND	11	92
	4b	462	17	ND	20	231
	15	139	28	14	28	3
	6a	37	6	ND	3	33
	6b	277	100	71	44	693
	18a	42	11	11	3	9
	18b	250	16	16	3	67

Structural formula	Compound	IS**				
		CC$_{50}$* (μM)	HSV-1	HSV-2	MV	JV
	8	901	32	32	2	3
	7a	819	63	63	28	31
	7b	1044	80	40	40	40
	31b	254	8	10	ND	ND
	10	935	11	11	1	65
	29b	322	11	22	ND	ND
	12a	100	3	1	1	3
	12b	160	109	27	54	27

Structural formula	Compound	IS**				
		CC$_{50}$* (μM)	HSV-1	HSV-2	MV	JV
	23	>890	10	10	ND	ND
	11	858	7	2	27	64
	24	520	7	<7	ND	ND
	32a	130	7	11	ND	ND
	32b	33	14	11	ND	ND
	Acyclovir (ACV)	280	933	467	--	--
	Ribavirin (RBV)	420	--	--	14	37

* 50% Cytotoxic concentration, or compound concentration required to reduce cell viability by 50% of the untreated control, after 24 h incubation at 37°C. Data are the average of duplicates.

** Selectivity index or ratio CC$_{50}$/EC$_{50}$. EC$_{50}$ values were determined by a virus yield reduction assay.

Data are the average of duplicates.

ND: not done.

Most of the compounds presented CC_{50} values higher than 100 μM except for **1a, 6a, 17b, 18a, 26a** and **32b** which showed CC_{50} values of 63, 37, 43, 42, <108 and 33 μM, respectively. This indicates a high degree of cytotoxicity for these analogues making them inappropriate as antiviral compounds. The rest of the compounds show an acceptable cytotoxicity with values close to that of stigmasterol [24, 25].

Stereochemistry of the side chain appears to play an important role in toxicity since most of the compounds with 22S, 23S configuration (**b** structure) are less toxic than those with 22R, 23R configuration (**a** structure) (see pairs **a-b** for compounds **1, 2, 4, 5, 6, 7, 12** and **18**); however, there are three compounds that do not follow this rule: **3, 17** and **32**.

The presence of an unsaturated function in the side chain (compounds **8, 10, 11, 23, 24** and **25**) increases CC_{50} values [25]; compounds **14** and **15** are clear exceptions.

The chemical protection of the hydroxyl groups on C-2, C-3 and C-22, C-23 by isopropylidene groups (compound **28a**), as well as the presence of a free C-3 hydroxyl group, substantially increase the CC_{50} values (compare **3a, 5a** and **17a** with **30a**, and **4a, 4b, 6a, 6b, 18a, 18b** with **7a** or **7b**).

The expansion of ring B, transforming the 6-keto group in a lactone one, decreases cytotoxicity, as shown on compounds **1a** and **9a**.

Inversion of the configuration of the hydroxyl group of C-3 on compounds **7b** and **30a** lead to a significant increase in cytotoxicity, as it is shown for compounds **31b** and **26a**, respectively. This effect does not occur in compounds bearing a C- fluorine group (see compounds **17a** and **22a**).

The presence of an additional hydroxyl group on C-5 does not follow a regular pattern; this would indicate that the presence of other functional groups contributes more significantly to the final toxicity. However, the presence of a fluorine group on C-5 highly increases the cytotoxicity, as compared to a hydroxyl group on the same position (see pairs **29b-2b, 12a-7a** and **12b-7b**) [25].

Additional steroidal-related compounds should be synthesized and further studies should be made to establish a more precise structure-activity relationship taking into account variations of cytotoxicity due to changes in the molecular structure.

ANTIVIRAL STUDIES

Throughout our studies, we have tried different methods to demonstrate that BRs inhibit virus replication. The data presented here were obtained by application of a virus yield reduction assay [21-25]. Briefly, Vero cells grown in 24 well culture plates, for 48 h, were infected with the assayed virus at a moi of 1. After 1 h of adsorption at 37°C, the cells were covered with maintenance medium (MM) containing various concentrations of the tested compounds. After 24 h of incubation at 37°C, the infected cultures were subjected to two cycles of freeze-thawing, followed by centrifugation at low speed (1000 x g). Then, the supernatants were titrated by a plaque assay. The EC_{50} (compound concentration required to reduce viral plaque formation by 50%) was determined for each compound.

Most of the assayed derivatives exhibit better selectivity indexes (SI) (ratio between CC_{50} and EC_{50}) than 28-homocastasterone (**1a**) which possesses the basic steroidal skeleton common to all of the synthetic compounds (Fig. **1**). Compounds **6b, 7b** and **12b** were active against all of the viruses assayed: HSV-1, HSV-2, MV, JUNV (Table **2**) and VSV (which present SI values of 41.7, 30 and 12.4, respectively, for the three assayed compounds). Table **3** shows the antiviral activity of some BRs at a concentration of 40 μM, against PV and VSV. The most active BRs, with inhibition percentages higher than 90 %, were **1b, 2b, 3b, 6b, 7b** and **12b**.

Table 3. Antiviral activity of BRs derivatives against VSV and PV.

Compound	Inhibition %	
	PV	VSV
1a	85	23
1b	98	100
2a	83	58
2b	99	92
3a	33	36
3b	100	94
4a	22	< 5
4b	48	<5
5a	64	<5
5b	72	<5
6a	82	46
6b	92	99
7b	90	92
12b	91	97

Vero cells were infected at a moi of 1 with PV or VSV. After 1 h adsorption at 37°C, the cultures were covered with MM or with MM containing the BR derivative at a concentration of 40 μM, except for compound 6a, which was assayed at 20 μM, because of its high cytotoxicity. After 24 h of incubation at 37°C, supernatants were harvested and titrated by a plaque assay.

Compound **6b,** was the most active against HSV-1 and HSV 2, although it presented lower activity than ACV [24, 25]. By contrast, compound **6b** showed 3 to 18-fold higher activity, when compared to ribavirin, against VSV, MV and JUNV, respectively [22, 25, 33]. BR-derivatives were highly effective to inhibit JUNV *in vitro* infection, with SI values higher than 100 for compounds **2a, 2b, 3b, 4b, 5a, 6b**. Similar results were obtained with TCRV and PICV, indicating that arenaviruses were very sensitive to the antiviral activity of this type of compounds [21]. Results obtained with the analogues were evaluated in order to establish a preliminary structure-activity relationship. Our findings indicate that analogues having the opposite unnatural configuration (22S, 23S) usually show enhanced activity (compare **4a** to **4b**, or **2a** to **2b**). On the other hand, introduction of an electronegative group at C-5 (fluorine in **12b** and hydroxyl in **7b**) leads to more active compounds against HSV-1.

None of the tested compounds exhibited direct inactivating effect on the virus particle, indicating that they inhibit a specific step of the virus multiplication cycle. Attempts to disclose the mode of action of **6b** against HSV-1 indicated that in the presence of the BR virus late protein synthesis was severely diminished (Fig. **3**) [24]. Since late protein synthesis is dependent on viral DNA replication, studies of drug-drug combination with inhibitors of viral DNA synthesis such as ACV and foscarnet (FOS) were performed. These assays demonstrated that **6b** would act synergistically with low concentrations of ACV and moderate concentrations of FOS against HSV [34]. These results suggest that the mechanism of antiviral action of **6b** differs from the antiviral mode of action of nucleoside analogues such as ACV.

The inhibitory action of **6b** against JUNV replication in Vero cells was also investigated [35]. Time of addition experiments revealed that **6b** was most effective the earlier it was added to the cells after infection, and that **6b** prevents the formation of mature viral particles rather than the release of progeny virus to the extracellular medium. Similarly to what had been previously found for HSV-1, neither adsorption nor internalization of viral particles was the target of the inhibitory action. We have also demonstrated that **6b** prevented early translation steps of the JUNV replication cycle (Fig. **3**) [35]. In addition, a high inhibition of virus yields and JUNV-mediated cell fusion was also observed when compound **6b** was present during the last steps of the virus multiplication cycle. Given these results, an adverse effect of **6b** on post-translational processing or proper insertion of JUNV glycoproteins into the cell membrane cannot be ruled out (Fig. **3**). The effect of **6b** on different steps of VSV replicative cycle was also investigated [33]. In this model, a high inhibition of viral protein synthesis and virus particle formation was evident and these results are in accord with experimental evidences obtained with JUNV (Fig. **3**). Although **6b** shows a broad antiviral spectrum, the fact that some viruses are more susceptible to its inhibitory effect than others indicates that **6b** could exert its antiviral action by affecting a specific viral factor and, to a lesser degree, a cellular function required for viral replication.

Mounting evidences point to the low level of toxicity of natural BRs. Being constituents of practically all plants, BRs are regularly consumed by mammals and the safety of some natural BRs has been confirmed by toxicological studies in mice and rats (orally and dermally) [36]. *In vivo* studies performed in a murine experimental model have demonstrated a non-toxic effect of **6b** (40 μM), after topical eye administration, when the compound was applied three times a day, during 3 consecutive days [37].

Evaluation of *in vivo* antiherpetic activity of synthetic BRs was performed using the murine herpetic stromal keratitis (HSK) experimental model. Administration of compounds **6b**, **7b** or **12b** to the eyes of mice at 1, 2 and 3 days post-infection delayed and reduced the incidence of HSK, however, viral titers of eye washes were not diminished in BR-treated mice, suggesting that the compounds do not exert a direct antiviral effect but rather they may play a role in immune-mediated stromal inflammation [37]. This hypothesis was further supported by *in vitro* studies that demonstrated that compound **6b** modulates the response of epithelial and immune cells to HSV-1 infection through the induction or inhibition of cytokine production, depending on the cell type involved [38]. Thus, the protective effect in mice could be due to a balance between the immunostimulating and immunosuppressive effects of the derivative. Similar results were obtained with other synthetic stigmastane analogues [39]. In addition, several synthetic stigmastanes were able to inhibit the TNF-α production in L929 cells. According to preliminary analysis of the structure–activity relationships, the modulatory effect on the TNF-α production of these synthetic compounds could be related with the presence of a hydroxylated stigmastane side chain, having a 22S, 23S configuration, and with a 3β, 5α-dihydroxy-6-keto moiety in the steroidal ring system as well [40].

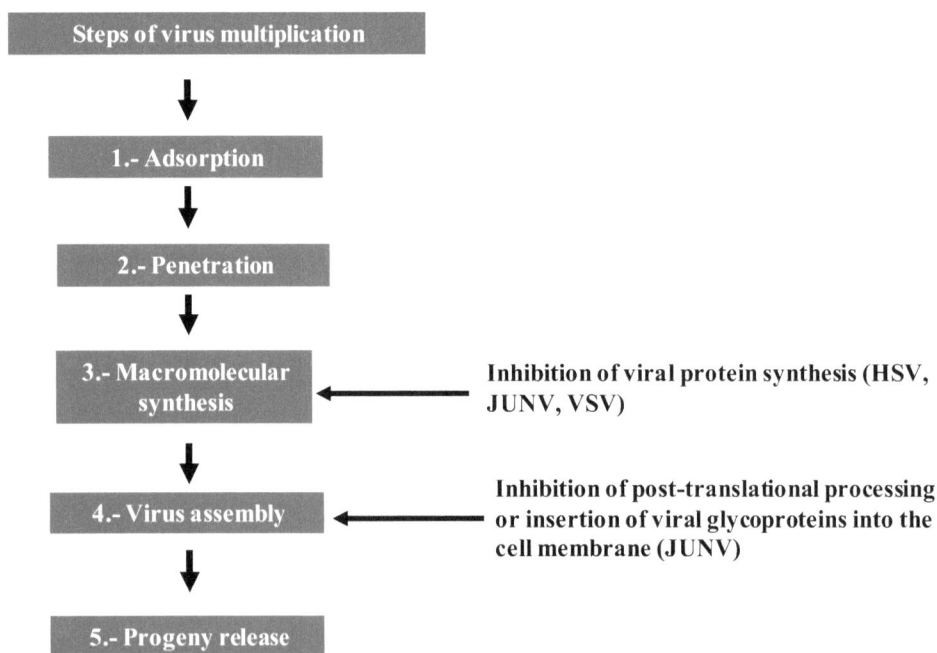

Figure 3: Possible antiviral targets of the BR **6b** in the virus multiplication cycle.

CONCLUSIONS

Although the role of sterols in cell biology has been widely explored, new functions for these molecules are continuously discovered [41, 42].

In this chapter, we describe the action of BR-like structures against different human pathogenic viruses. We have found that most of the tested compounds are active against HSV-1, HSV-2, MV, PV, VSV and JUNV, at concentrations that were not toxic to the host cells. Three of them: **6b** [(22S,23S)-3β-bromo-5α,22,23-trihydroxystigmastan-6-one], **7b** [(22S,23S)-3β,5α,22,23-tetrahydroxystigmastan-6-one] and **12b** [(22S,23S)-5α-fluoro-3β,22,23-trihydroxystigmastan-6-one] showed significant selectivity indexes against all assayed viruses.

Most of the BRs showed bioactivity 10 to 100-fold higher than the bioactivity reported for other natural steroidal agents [10-20, 26], however, besides a specific inhibition of a viral multiplication step, we can not rule out that BRs would also exert an indirect antiviral action by affecting a cellular function required for viral replication. It is known, that BRs function as signaling molecules in plants and that they are perceived at the plasma membrane by direct binding to the extracellular domain of a transmembrane protein, BRI1, which has serine-threonine protein kinase activity. BR binding initiates a signaling cascade and further kinases and phosphatases determine the phosphorylation state and stability of nuclear transcription factors. These factors mediate major BR effects in various plant physiological processes [43]. Therefore, it would be interesting to study the effect of BRs on components of the animal cell signal transduction system, such as MAPKs (mitogen-activated protein kinases). It has been reported that certain steroidal compounds, like DHEA, induces a p38 MAPK-phosphatase that suppress the p38-MAPK cascade [44]. Controversial results have been reported about the action of DHEA on ERK (extracellular stimulus-regulated kinases) MAPK pathway, in different cell systems. In certain cell lines, DHEA induces ERK phosphorylation [20], whereas in other cell types DHEA treatment inhibits ERK activation [45]. Several viruses, like AdV, not only induce ERK phosphorylation but also seem to depend on it for its multiplication. It was also proved that the level of viral protein synthesis is strongly reduced by inhibitors of ERK activation [46], so, a possible explanation for the antiviral activity of BRs may be related to the ability of these steroids to modulate MAPK activation or inhibition induced by virus infection. Such actions could explain, at least partially, the broad antiviral spectrum of this kind of compounds.

In summary, natural and synthetic BRs evaluated in this study represent an important class of compounds shown to be active against several RNA and DNA viruses. However, further synthetic modifications, based on structure activity relationship studies, should be done to decide if the BRs family may lead to a new class of viral inhibitors. The mechanism underlying the antiviral mode of action of this kind of compounds and its relation with host cell signal transduction system must also be further explored.

ACKNOWLEDGEMENTS

We thank Dr. Lydia R. Galagovsky and Dr. Javier A. Ramirez for the synthesis of the brassinosteroid derivatives. This work was supported by grants from the Universidad de Buenos Aires, UBA, UBACYT X 505 and Agencia Nacional de Promoción Científica y Técnica (ANPCYT) PICT 00985/07.

REFERENCES

[1]   Elion, GB. Mechanism of action and selectivity of acyclovir. Am J Med 1982; 73, (Suppl. 1A), 7-13.

[2]   De Clercq E. Highlights in the discovery of antiviral drugs: A personal retrospective. J Med Chem 2010; 53: 1438-50.

[3]   De Clercq E. Anti-HIV drugs: 25 compounds approved within 25 years after the discovery of HIV. Int J Antimicrob Agents 2009; 33: 307-20.

[4]   Treanor J, Falsey A. Respiratory viral infections in the elderly. Antiviral Res 1999; 44: 79-102.

[5]   Shigeta S. Approaches to antiviral chemotherapy for acute respiratory infections. Antiviral Chem Chemother 1998; 9: 93-107.

[6]   Dixon, R. A. Natural products and plant disease resistance. Nature 2001; 411: 843-7.

[7]   Grove MD, Spencer GF, Rohwedder WK, *et al.* Brassinolide, a plant growth-promoting steroid isolated from *Brassica napus* pollen. Nature 1979; 281: 216-7.

[8]   Mandava NB. Plant growth-promoting brassinosteroids. Ann Rev Plant Physiol Plant Mol Biol 1988; 39: 23-52.

[9]   Bajguz A, Tretyn A. The chemical characteristic and distribution of brassinosteroids in plants. Phytochem 2003; 62: 1027-46.

[10] Kohen F F, Gunasekera M, Cross S S. New antiviral sterol disulfate ortho esters from the marine sponge Petrosia weinbergi. J Org Chem 1991; 56: 1322–5.

[11] Sun H H, Cross S S, Gunasekera M, Koehn F F. Weinbersterol disulfates A and B, antiviral steroid sulfates from the sponge *Petrosia weinbergi*. Tetrahedron 1991; 47: 1185-90.

[12] Comin M J, Maier M S, Roccatagliata A J, Pujol C A, Damonte, E B. Evaluation of the antiviral activity of natural sulfated polyhydroxysteroids and their synthetic derivatives and analogs. Steroids 1999; 64: 335-40.

[13] Arthan D, Svasti J, Kittakoop P, Pittayakhachonwut D, Tanticharoen M, Thebtaranonth Y. Antiviral isoflavonoid sulfate and steroidal glycosides from the fruits of *Solanum torvum*. Phytochem 2002; 59: 459-63.

[14] Acosta E G, Bruttomesso A C, Bisceglia J A, Wachsman M B, Galagovsky L R, Castilla V. Dehydroepiandrosterone, epiandrosterone and synthetic derivatives inhibit Junin virus replication *in vitro*. Virus Res 2008; 135: 203-12.

[15] Romanutti C, Bruttomesso A C, Castilla V, Bisceglia J A, Galagovsky L R, Wachsman M B. *In vitro* antiviral activity of dehydroepiandrosterone and its synthetic derivatives against vesicular stomatitis virus. Vet J 2009; 182:327-35.

[16] Romanutti C, Bruttomesso A C, Castilla V, Galagovsky L R, Wachsman M B. Anti-adenovirus activity of epiandrosterone and dehydroepiandrosterone derivatives. Chemotherapy 2010; 56:158-65.

[17] Henderson E, Yang J, Schwartz A: Dehydroepiandrosterone (DHEA) and synthetic DHEA analogs are modest inhibitors of HIV-1 IIIB replication. AIDS Res Human Retrovir 1992; 8: 625-31.

[18] Diallo K, Loemba H, Oliveira M, Mavoungou D D, Wainberg M A. Inhibition of human immunodeficiency virus type-1 (HIV-1) replication by immunor (IM28), a new analog of dehydroepiandrosterone. Nucleos Nucleot Nucleic Acids 2000; 19: 2019-24.

[19] Mavoungou D, Poaty-Mavoungou V, Akoume M, Ongali B, Mavoungou E: Inhibition of human immunodeficiency virus type-1 (HIV-1) glycoprotein-mediated cell-cell fusion by immunor (IM28). Virol J 2005; 11: 9-10.

[20] Chang CC, Ou YC, Rang SL, Chen CJ. Antiviral effect of dehydroepiandrosterone on Japonese Encephalitis virus infection. J Gen Virol 2005; 86: 2513-23.

[21] Wachsman MB, López EM, Ramírez, JA, Galagovsky LR, Coto CE. Antiviral effect of brassinosteroids against herpes virus and arenaviruses. Antivir Chem Chemother 2000; 11: 71-7.

[22] Wachsman MB, Ramirez JA, Galagovsky LR, Coto CE Antiviral activity of brassinosteroids derivatives against measles virus in cell cultures. Antivir Chem Chemother 2002; 13: 61–6.

[23] Talarico LB, Ramirez JA, Galagovsky LR, Wachsman MB. Structure–activity relationship studies in a set of new brassinosteroid derivatives assayed against herpes simplex virus type 1 and 2 in cell cultures. Med Chem Res 2002, 11: 434–44.

[24] Wachsman MB, Castilla V, Talarico LB, Ramirez JA, Galagovsky LR, Coto CE. Antiherpetic mode of action of (22S,23S)-3β-bromo-5α,22,23-trihydroxystigmastan-6-one *in vitro*. Int J Antimicrob Agents 2004; 23: 524-7.

[25] Wachsman MB, Ramírez JA, Talarico LB, Galagovsky LR, Coto CE. Antiviral activity of natural and synthetic brassinosteroids. Curr Med Chem.- Anti-Infective Agents 2004; 3: 163-79.

[26] Castilla V, Ramírez JA, Coto CE. Plant and animal steroids a new hope to search for antiviral agents. Curr Med Chem 2010; 17: 1858-73.

[27] Cutts FT, Henao-Restrepo AM, Olive JM. Measles elimination: progress and challenges. Vaccine 1999; 17: S47-S52.

[28] Weissenbacher MC, Laguens RP, Coto CE. Argentine hemorrhagic fever. Curr Top Microbiol Immunol 1987; 134: 79-116.

[29] Collett MS, Neyts J, Modlin JF. A case for developing antiviral drugs against polio. Antivir Res 2008; 79:179-87.

[30] Rodriguez LL. Emergence and re-emergence of vesicular stomatitis in the United States. Virus Res 2002; 85: 211-9.

[31] Whitley RJ, Roizman B. Herpes simplex virus infections. The Lancet 2001; 357: 1513-8.

[32] Denizot F, Lang R. Rapid colorimetric assay for cell growth and survival. J Immunol Methods 1986; 89: 271-7.

[33] Romanutti C, Castilla V, Coto CE, Wachsman MB. Antiviral effect of a synthetic brassinosteroid on the replication of vesicular stomatitis virus in Vero cells. Int J Antimicrob Agents 2007; 29: 311–6.

[34] Talarico LB, Castilla V, Ramirez JA, Galagovsky LR, Wachsman M B. Synergistic *in vitro* interactions between (22S,23S)-3β-bromo-5α,22,23-trihydroxystigmastan-6-one and acyclovir or foscarnet against herpes simplex virus type 1. Chemotherapy 2006; 52: 38-42.

[35] Castilla V, Larzábal M, Sgalippa NA, Wachsman M B, Coto C E. Antiviral mode of action of a synthetic brassinosteroid against Junin virus replication. Antiviral Res 2005; 68: 88-95.

[36] Khripach V, Zhabinskii V, De Groot A. Twenty years of brassinosteroids: Steroidal plant hormones warrant better crops for the XXI century. Ann Bot 2000; 86: 441-7.

[37] Michelini F M, Ramírez JA, Berra A, Galagovsky LR, Alché L E. *In vitro* and *in vivo* antiherpetic activity of three new synthetic brassinosteroid analogues. Steroids 2004; 69: 713-20.

[38] Michelini FM, Berra A, Alché LE. The *in vitro* immunomodulatory activity of a synthetic brassinosteroid analogue would account for the improvement of herpetic stromal keratitis in mice. J Steroid Biochem Mol Biol 2008; 108: 164-70.

[39] Michelini FM, Ramírez JA, Berra A, Galagovsky LR, Alché L E. Anti-herpetic and anti-inflammatory activities of two new synthetic 22,23-dihydroxylated stigmastane derivatives. J. Steroid Biochem Mol Biol 2008; 111: 111-6.

[40] Ramírez JA, Bruttomesso AC, Michelini FM, Acebedo SL, Alché LE, Galagovsky LR. Syntheses of immunomodulating androstanes and stigmastanes: comparison of their TNF-alpha inhibitory activity. Bioorg Med Chem 2007; 15: 7538-44.

[41] Mauch DH, Nagler K, Schumacher S, *et al.* CNS synaptogenesis promoted by glia-derived cholesterol. Science 2001; 294: 1354–7.

[42] Schaller H. The role of sterols in plant growth and development. Prog Lipid Res 2003: 42:163-75.

[43] Karlova R, de Vries S C. Advances in understanding brassinosteroid signaling. Sci STKE; 2006: 354 pe36.

[44] Ashida K, Goto K, Zhao Y, *et al.* Dehydroepiandrosterone negatively regulates the p38 mitogen-activated protein kinase pathway by a novel mitogen-activated protein kinase phosphatase. Biochim Biophys Acta 2005; 1728:84-94.

[45] Ziegler CG, Sicard F, Sperber S, Ehrhart-Bornstein M, Bornstein S R, Krug A W. DHEA reduces NGF-mediated cell survival in serum-deprived PC12 cells. Ann N Y Acad Sci 2006; 1073: 306-11.

[46] Schümann M, Dobbelstein M. Adenovirus induced extracellular signal-regulated kinase phosphorylation during the late phase of infection enhances viral protein levels and virus progeny. Cancer Res 2006; 66: 1282-8.

CHAPTER 7

Antiherpetic and Anti-Inflammatory Activities of Novel Synthetic Brassinosteroids Analogs

Laura E. Alché* and Flavia M. Michelini

Laboratorio de Virología: Agentes antivirales y citoprotectores. Departamento de Química Biológica. Facultad de Ciencias Exactas y Naturales, Universidad de Buenos Aires, Ciudad Universitaria, Pabellón 2, Piso 4, C1428EGA, Buenos Aires, Argentina

Abstract: Many viral infections are associated with the development of immunopathologies and autoimmune diseases, of difficult treatment, for which no vaccines are available yet. Obtaining compounds that conjugate both antiviral and immunomodulating activities in the same molecule would be very useful for the prevention and/or treatment of these immunopathologies of viral origin. Within this chapter, we present the evaluation of biological properties of synthetic brassinosteroids (BRs) analogs, and stigmastane and androstane derivatives, as potential antiviral and immunomodulating agents. We chose three synthetic BRs which had exhibited high antiviral activity against different human pathogenic viruses. The new steroidal molecules were designed with chemical modifications from the three BRs, in order to improve their antiviral activity. We also included some structural moieties responsible for the immunommodulating activity of well-known anti-inflammatory steroids.

Keywords: Conjunctival cell lines, herpes simplex virus, herpetic stromal keratitis, human corneal cell line.

INTRODUCTION

Pathogenesis of viral infections lye on the occasional damage produced by the virus on the cells that it infects which leads to cell and tissue disfunction in the host [1]. Virus may cause cellular damage in different ways. Some viruses produce a direct lesion on the infected tissue due to an excessive production of viral progeny, or the destabilization of cellular membrane during viral multiplication. Viruses responsible for respiratory and gastrointestinal tract infections, such as the Ebola virus and Herpesviruses, have been associated to this kind of damage, in which cell death is a consequence of viral cytopathic effect [2, 3].

Cell death of infected tissues may also occur by apoptosis induced by the virus itself, in order to favor its own liberation and propagation, or by the infected cell, as a part of the innate or natural defense mechanism to reduce viral load and dissemination [4].

Some viruses produce cell damage through cell transformation or malignization caused by modification of the host cell genome or desregulation of the cell cycle. Examples of these oncogenic viruses are Retroviruses, Human Papillomaviruses, Epstein-Barr and Hepatitis B viruses [5, 6].

Many viruses that cause disease directly or by transformation of infected cells also possess the ability to induce an immune response that contributes to viral disease, hence, the disease they provoke derives from more than one pathogenic mechanism. Most of these viruses are of public health significance because they cause damage through triggering an immunopathology. This means that the damage is produced by the immune response or host inflammatory response against the virus. Since there are neither vaccines nor antiviral compounds available to prevent or treat many of these diseases, common treatment for them consists on the administration of immunosuppressive drugs. However, they may provoke viral reactivation and a prolonged illness, as it occurs in the case of myocarditis induced by Coxsackie B virus, and Herpetic Stromal Keratitis (HSK), due to Herpes Simplex Virus (HSV) [7, 8].

*Address correspondence to Laura E. Alché: Laboratorio de Virología: Agentes antivirales y citoprotectores. Departamento de Química Biológica. Facultad de Ciencias Exactas y Naturales, Universidad de Buenos Aires, Ciudad Universitaria, Pabellón 2, Piso 4, C1428EGA, Buenos Aires, Argentina. E-mail: lalche@qb.fcen.uba.ar

Adaucto Bellarmino Pereira-Netto (Ed)

Besides producing ulcerative lesions on mucous membranes, HSV-1 can trigger an immunopathology when the site of infection is the eye, starting the illness known as HSK. In this case, the inflammatory response elicited to remove the virus causes irreversible damage of the cornea, damage that seriously compromises the host vision and even may cause blindness. Furthermore, due to the HSV ability to establish latency, recurrent episodes of quiescent virus reappearing in the cornea from the innervating ganglion are a significant cause of ocular morbidity [9-12]. Therefore, HSK is an immunopathology of viral origin in which the lesions observed in the eye are not caused by viral replication *per se*, but as a consequence of the inflammatory response triggered by viral replication. Ocular lesions become evident when the virus is no longer detectable in the eye.

Human HSK is not healed by antivirals, but its symptoms can be mitigated by immunosupressive agents, including topical or systemic corticosteroids and Cyclosporine A. Studies carried out on therapeutic vaccination demonstrated that this technique would be an efficient way to reduce recurrent HSV ocular disease by inhibiting viral reactivation in latently infected ganglia. Another approach to overcome corneal opacification due to HSK is the keratoplasty (corneal transplantation) [12, 13]. However, HSK treatment also includes the use of acyclovir (**ACV**), the antiviral drug of choice to treat HSV infections, to diminish an eventual viral reactivation due to the immunosupression provoked by both drugs [13]. Although the condition of some patients with chronic HSK can be stabilized by intense topical and systemic antiviral and anti-inflammatory drug regimens, many of them do not respond to therapy: both, the increased neovascularization and the use of steroids augment the risk of corneal graft failure [14].

Therefore, it is of great interest to have compounds with both antiviral and anti-inflammatory activities that inhibit viral multiplication and/or the progression of the immune response to infection.

From the antiviral activity studies performed with synthetic BR analogs, it was observed that compounds developed by Ramírez *et al.* [15] (please see Chapter 6) [16] such as (*22S,23S*)-3β-bromo-5α,22,23-trihydroxystigmastan-6-one (**6b**), (*22S,23S*)-3β,5α,22,23-tetrahydroxystigmastan-6-one (**7b**) and (*22S,23S*)-5α-fluoro-3β,22,23-trihydroxystigmastan-6-one (**12b**) exhibit the highest biological activity to inhibit HSV-1 multiplication in Vero cells (please see Table **2** in Chapter 6) [17 - 21]. Considering the promising results obtained with these BR analogs, grouped arbitrarily as Group I (Fig. **1**), we decided to synthesize new steroidal molecules with chemical modifications aimed at enhancing their antiviral activity and acquiring immunomodulating properties, as well.

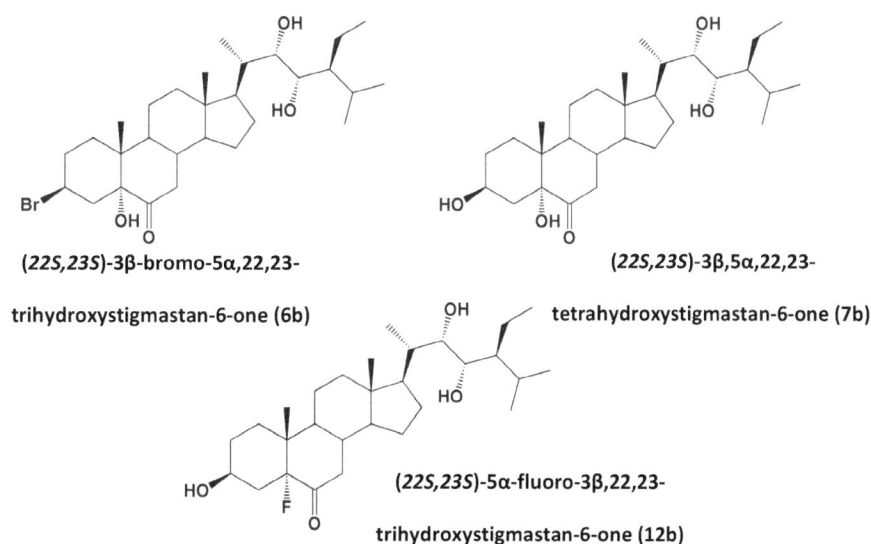

(*22S,23S*)-3β-bromo-5α,22,23-

trihydroxystigmastan-6-one (6b)

(*22S,23S*)-3β,5α,22,23-

tetrahydroxystigmastan-6-one (7b)

(*22S,23S*)-5α-fluoro-3β,22,23-

trihydroxystigmastan-6-one (12b)

Figure 1: Chemical structures of synthetic analogs **6b**, **7b** y **12b** (Group I).

Synthesis of novel BRs analogs

Given the importance that finding compounds blending together antiviral and anti-inflammatory activities could have for the treatment of viral immunopathologies, we decided to synthesize new molecules in which structural features present in well-known anti-inflammatory steroids were introduced.

Hence, three new families of compounds were designed and grouped into Groups II, III and IV; the synthesis of compounds belonging to Groups II and III was carried out starting from stigmasterol, while androstanes from Group IV were obtained from dehidroepiandrosterone (**DHEA**).

Group II molecules synthesis (Fig. **2A**) was directed to obtain compounds that, keeping the dihydroxylated side chain responsible for the antiviral activity of BRs, also showed structural similarity to dexamethasone (**DEX**) in the A ring of the steroid (**F11**) (Fig. **2B**). This group included compound **32b** (please see Table **2** in Chapter 6), since the A ring was similar to that of cortisone (Fig. **2B**).

DHEA and its sulphate esther (**S-DHEA**) are the most abundant circulating steroid hormones in human beings and possess potentially beneficial effects, such as immunomodulatory and antiviral activities [22 - 24]. Therefore, the study of new molecules structurally related to **DHEA** would be useful in the search for compounds with the advantageous properties of this steroid. In this sense, synthesis of the compounds corresponding to Groups III and IV were performed. Whereas molecules from Group III kept the dihydroxylated side chain of the stigmastane derivatives and the modifications were done in rings A and B of the steroid (Fig. **3**),

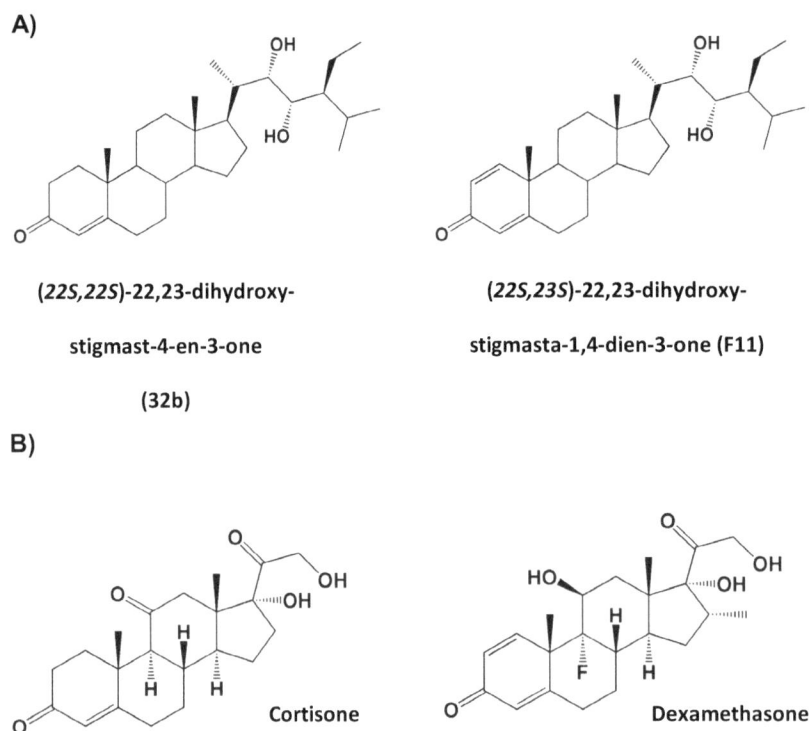

A)

(22S,22S)-22,23-dihydroxy-

stigmast-4-en-3-one

(32b)

(22S,23S)-22,23-dihydroxy-

stigmasta-1,4-dien-3-one (F11)

B)

Cortisone

Dexamethasone

Figure 2: Structures of **A)** synthetic stigmastanes **32b** and **F11** and **B)** steroids cortisone and **DEX**.

Group IV consisted of compounds that maintained the same substitutions in A and B rings of Group I, without the dihydroxylated side chain of the BRs (Fig. **4**) [25].

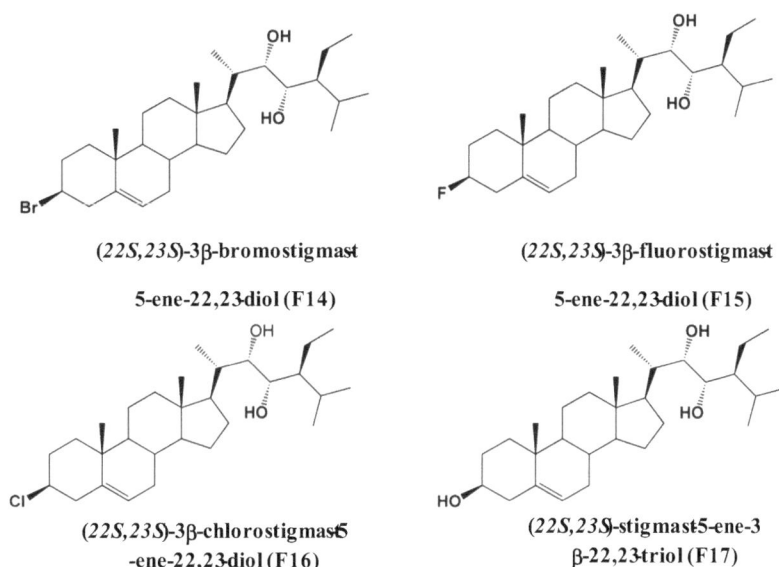

(*22S,23S*)-3β-bromostigmast

5-ene-22,23-diol (F14)

(*22S,23S*)-3β-fluorostigmast

5-ene-22,23-diol (F15)

(*22S,23S*)-3β-chlorostigmast-5
-ene-22,23-diol (F16)

(*22S,23S*)-stigmast-5-ene-3
β-22,23-triol (F17)

Figure 3: Structures of synthetic stigmastanes **F14-F17**.

IN VITRO ANTIVIRAL ACTIVITY OF STIGMASTANE/ANDROSTANE ANALOGS

The ocular surface is formed by the epithelium of the cornea and conjunctiva, which show a close histological continuity and linked functionality [26]. Although the corneal epithelium represents the main substrate for HSV-1 multiplication, the conjunctiva is a natural target for the virus, and its continuity with the corneal epithelium allows viral spread [27]. Besides, it has been observed that the severity of the signs of primary ocular infection in the conjunctiva correlates better with a higher incidence of recurrent infection than the severity of the signs in corneal infection does [28]. Hence, both types of cells are involved in HSK development [29]. Therefore, it is of great interest to evaluate the effect of antiviral compounds in ocular epithelial cells, with the final purpose of healing HSK.

In order to study the inhibitory properties of the synthetic stigmastanes and androstanes against HSV-1, we determined their cytotoxicity and antiviral activity *in vitro*, in human corneal (HCLE) and conjunctival (IOBA-NHC) cell lines. Cytotoxicity was determined by the colorimetric MTT assay [30] to establish the concentration of each compound that reduced 50% of cell viability (CC_{50}). The CC_{50} values of each compound were similar in both cellular types (Table **1**).

BRs **6b** and **12b** and compounds from Groups III and IV did not exhibit toxicity, even at the maximum concentrations tested, while BR **7b** and stigmastanes from Group II showed the lowest CC_{50} values, indicating their highest cytotoxicity.

By means of viral yield reduction assays, the concentration of each compound needed to reduce 50% of the viral yield (EC_{50}) was established and, thus, the antiviral activity. The most active compounds were BRs **6b**, **7b** and **12b** (Group I) and the stigmastane analog **32b** (Group II) which showed the highest selectivity indices (SI) (CC_{50}/EC_{50}) (Table **1**).

After infection of IOBA-NHC cells, BRs **6b**, **7b** and **12b** proved to inhibit HSV-1 yield with SI of >5540, 768.4 and 490.6, respectively (Table **1**) [16]. The mild anti-HSV-1 activity of compound **32b** observed could be ascribed to the presence of the Δ^4-3-ceto moiety (Fig. **2A**). The extra double bond Δ^1 in **F11** seems to reduce this activity even more (Fig. **2A**).

None of the androstane analogs displayed anti-HSV-1 activity, and only one stigmastane from Group III (**F16**) showed a SI higher than 10 in IOBA-NHC cells (Table **1**). The removal of the dihydroxylated side chain of androstane analogs completely abolished their antiviral activity (Table **1**, Fig. **4**), which confirmed the results already reported by Wachsman *et al.*, relative to the contribution of the side chain to the antiviral activity of BRs [21]. Although compounds from Group III kept the dihydroxylated side chain, the substitution of the 5α-hydroxy-6-keto structure in BRs **6b** and **7b** by the Δ^5 double bond, to render the corresponding derivatives **F14** and **F17** (Figs. **1** and **3**), resulted in a marked drop in the antiviral activity (Table **1**). The ability of compounds to inhibit viral multiplication was less evident in corneal cells, since only BR **6b** exhibited a SI higher than 10 (Table **1**).

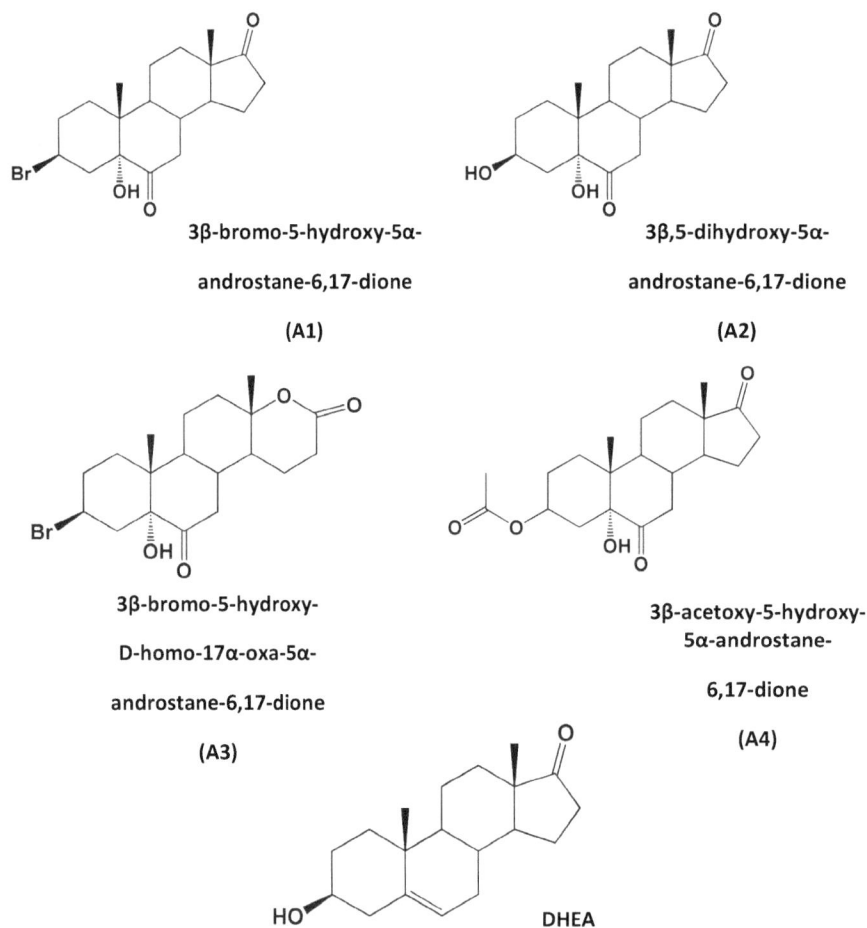

3β-bromo-5-hydroxy-5α-androstane-6,17-dione

(A1)

3β,5-dihydroxy-5α-androstane-6,17-dione

(A2)

3β-bromo-5-hydroxy-D-homo-17α-oxa-5α-androstane-6,17-dione

(A3)

3β-acetoxy-5-hydroxy-5α-androstane-6,17-dione

(A4)

DHEA

Figure 4: Structure of synthetic androstanes **A1-A4** (Grupo IV) and **DHEA**.

Considering that the emergence of HSV-1 mutants resistant to antivirals commercially available is a frequent event, we evaluated the inhibitory effect of BR **6b** against **ACV** resistant HSV-1 (TK⁻) multiplication in HCLE cells. Fields and B2006 strains of TK⁻ viral mutants displayed the same replication efficiency as the wild type (wt) KOS strain in HCLE cells did (Table **2**). When HCLE infected cells were treated with 7 μM of **ACV**, more than two logarithm of reduction in HSV-1 wt yield was observed, whereas it failed to inhibit TK⁻ strains multiplication: a 10 fold higher dose was required to achieve less than one logarithm decrease in virus yield (Table **2**).

A concentration of 50 μM of BR **6b** was needed to reduce wt HSV-1 yield to the same extent as **ACV** (7 μM) did, and, surprisingly, BR **6b** was also efficient to inhibit **ACV** resistant TK- strains (Table **2**). These results are in agreement with data previously published by Wachsman *et al.* [18].

Table 1. Cytotoxicity and anti HSV-1 activity of the synthetic stigmastans and androstanes in NHC and HCLE cells [a] Cytotoxic concentration 50 %: compound concentration that reduces 50 % cell viability, determined by the colorimetric MTT assay in NHC and HCLE cells after 24 h incubation in the presence of each compoound. [b] Effective concentration 50 %: compound concentration required to reduce 50 % of viral yield in NHC y HCLE cells. [c] Selectivity Index: CC_{50}/EC_{50}. nd: not determined in: inactive

Group	Compound	MW	NHC			IICLE		
			CC_{50} (μM)	EC_{50} (μM)	SI	CC_{50} (μM)	EC_{50} (μM)	SI
I	**6b**	541	> 277	0.05 ± 0.01	> 5540	> 277	15.7 ± 0.9	> 17.6
	7b	478	146 ± 1.6	0.19 ± 0.08	768.4	226.4 ± 5.3	43.5 ± 1.2	5.2
	12b	480	> 314	0.64 ± 0.23	> 490.6	> 314	72.3 ± 0.5	4.3
II	**32b**	444	71.2 ± 16.7	5.4 ± 1.6	13.2	94.1 ± 0.8	10.1 ± 0.1	9.3
	F11	442	70.8 ± 1.7	17.9 ± 0.4	4	50 ± 0.4	in	in
III	**F14**	510	> 500	100	> 5	> 500	in	in
	F15	449	> 500	in	in	> 500	in	in
	F16	465.5	> 500	40	> 12.5	> 500	in	in
	F17	447	> 500	in	In	> 500	in	in
IV	**A1**	383	> 800	in	In	> 800	in	in
	A2	320	604.2 ± 99.5	in	in	600 ± 2.5	in	in
	A3	399	> 800	in	in	> 800	in	in
	A4	378	> 800	in	in	> 800	in	in

Table 2. Antiviral activity of BR **6b** against **ACV** resistant Fields and B2006 strains. of HSV-1

	CV	ACV		6b
		7 µM	70 µM	50 µM
KOS	5×10^7	1.7×10^5	n.d.	3.1×10^5
Fields	3.1×10^7	3×10^7	6.5×10^5	8×10^4
B2006	2.9×10^7	1.2×10^7	7.7×10^5	8×10^4

Virus yield expressed in PFU/ml.

n.d.: not determined

IN VITRO IMMUNOMODULATING ACTIVITY OF STIGMASTANE/ANDROSTANE ANALOGS

Macrophages are ubiquitous cells from the phagocytic mononuclear system. They exert different functional abilities like phagocytosis, microbian death, motility and surface adherence [31]. Besides, these cells play an important role in non-specific defense against viral infections [32, 33]. For example, they have a high intrinsic anti-HSV-1 activity by being non-permissive to viral replication and impeding viral spread [34].

Macrophages participate in viral clearance and also act as a source of inflammatory mediators, like cytokines. In the case of HSK, although macrophages are initially involved in viral clearance, they intervene in the progression of the immunopathology later [9]. These cells are not present in the healthy cornea, but they have been detected in this area after experimental HSV-1 corneal infection in mice [35-37].

The contribution of macrophages to the inflammatory response occurs through the secretion of pro-inflammatory molecules, like TNF-α [9]. To determine the eventual immunomodulating activity of synthetic compounds, we studied their effect on the secretory activity of TNF-α in macrophages. We determined the ability of compounds to inhibit TNF-α production in J774A.1 cells induced with LPS, using **DEX** as a positive control of inhibitory activity of TNF-α production [38]. **DHEA** was also included as a control in order to compare its activity with the activity observed for the compounds from Group IV, with structural similarity [39 - 41]. We found that **DEX** showed higher efficiency, when compared to **DHEA**, to restrain TNF-α secretion in activated macrophages (Table **3**).

BRs **6b** and **12b** were not toxic for J774A.1 cells, even at the highest concentrations tested (Table **3**). Except for compounds **F17** and **A2**, stigmastanes and androstanes from Groups III and IV also showed high CC_{50} values (Table **3**). It is interesting to highlight the structural similarity between compounds **F17** and BR **7b**, which not only share the dihydroxylated side chain, but also have hydroxyl groups in the A ring, and exhibited the smaller CC_{50} (Table **3**) (Figs. **1** and **3**). **A2** presents hydroxyl groups in rings A and B too, like BR **7b** and, in fact, it turned out to be more toxic, compared to the remaining compounds (Figs. **1** and **4**, Table **3**). Surprisingly, stigmastanes **32b** and **F11**, from Group II, were not toxic to J774A.1 cells (Table **3**). Stigmastanes **32b** and **F11**, from Group II, were the most active to impede TNF-α secretion, with the lowest IC_{50} values (Table **3**). These stigmastanes share structural similarity with anti-inflammatory corticosteroids, in the ring A, and were more effective than **DEX** (Table **3**, Fig. **2**).

With respect to Group I, BR **6b** and **7b** had IC_{50} values similar to that of **DEX**, and lower than that of **DHEA**. Nevertheless, **7b** exhibited a CC_{50} value close to the IC_{50}, which diminished its activity range with respect to **DHEA** and **DEX** controls, and BR **6b**.

The substitution of the hydroxyl group in C5 for a fluoride atom (F) in the **7b** molecule gave rise to a BR with no inhibitory activity of TNF-α production (**12b**) (Table **3** and Fig. **1**).

Compound **F17**, from Group III, had an IC_{50} value lower than the IC_{50} value found for **DHEA**, although its CC_{50} was low, which reduced its effectiveness. The substitution with halogens in C3 did not enhance the activity of compound **F17**.

In the case of compounds from Group IV, the substitution of the side chain of the stigmastanes by the keto group provoked a decrease in anti TNF-α activity. This is observed in the case of compounds **A1** and **A2**, that exhibited a lower inhibitory effect than their stigmastane analogs **6b** and **7b**, respectively (Figs. **1** and **4**). Oxidation of the keto group in C17 to render a lactone ring in androstane **A3** reduced the activity even more. Compound **A4** was the only androstane (Group IV) which presented an IC_{50} value similar to that of **6b**, and lower than **DHEA**. Compounds with a 5α-hydroxi-6-keto in rings A and B (**6b** and **7b**) were more active than the corresponding derivatives with a double bond Δ^5 (**F14** and **F17**) (Table **3**, Figs. **1** and **3**).

Table 3. *In vitro* inhibitory activity of TNF-α secretion in J774A.1 cells.

Group	Compound	MW	J774A.1	
			$CC_{50}{}^a$ (µM)	$CI_{50}{}^b$ (µM)
I	6b	541	> 277	38
	7b	478	67.9 ± 1.3	18
	12b	480	> 314	in
II	32b	444	> 314	1.6
	F11	442	> 314	5.6
III	F14	510	> 500	144
	F15	449	> 500	In
	F16	465.5	> 500	In
	F17	447	41.2 ± 0.8	35
IV	A1	383	> 800	134
	A2	399	366.6 ± 2.2	109
	A3	320	> 800	322.4
	A4	378	> 800	46.2
Controls	DHEA	288	> 3467	141
	DEX	392.5	> 3000	48

[a]50 % Cytotoxic Concentration: compound concentration that reduces 50 % of cell viability, determined in J774A.1 cells, by MTT method, after 24 h incubation with the compounds.

[b]50 % Inhibitory Concentration: compound concentration that reduces 50 % of TNF-α production in J774A.1 cells stimulated with 100 ng/ml LPS for 8 h.
in: inactive

STIGMASTANE ANALOGS AS PROMISING DRUGS TO HEAL HSK

Human necrotizing stromal keratitis (HSK) may be reproduced by the ocular infection of mice with HSV-1. In this experimental model, infection of the cornea initiates a series of events that culminate, after one or two weeks, in a disease hystologically similar to the HSK [42]. Moreover, it generally resembles many features of recurrent disease in humans, both, in the duration of viral shedding and in the effects of antiviral drugs [43].

By the use of the murine ocular model of infection, many advances have been achieved on the study of the pathogenic mechanisms that develop in the course of HSK, as well as the early events prior to the development of lesions, which set the stage for the subsequent pathology. These include viral replication, many cytokines production and corneal neo-vascularization which facilitate the arrival of the inflammatory cells responsible for the immunopathology [9]. The ocular infection induced by HSV in mice represents a suitable alternative for testing antiviral and anti-inflammatory drugs *in vivo* [44, 45], once it reproduces a pathology that meets both components, viral and inflammatory, of an immunopathology originated by HSV-1 infection.

During the first 5 days post-infection (p.i.), the virus replicates in the cornea, event that can be determined by the quantification of the viral load in ocular washes. In this first stage of disease, it is possible to evaluate the antiviral effect of a compound on the multiplication of HSV-1. Determination of infectious virus in the washes obtained from infected-treated eyes allow us to determine whether or not there is a decrease in viral titers, compared to untreated infected controls.

From day 5 p.i. onwards, when viral titers begin to decline in the eye, the first signs of the immunopathology appear. When infectious virus is no longer detected in the eyes, and the signs of the disease start to show up, the experimental model is useful to test a potential anti-inflammatory activity of a new molecule.

BRs **6b**, **7b** and **12b** and stigmastane analog **32b** were the most active antiviral compounds *in vitro* [16, 46]. Hence, they were assayed in the HSV-1-induced ocular infection in mice; neither of these stigmastanes had a toxic effect *in vivo*.

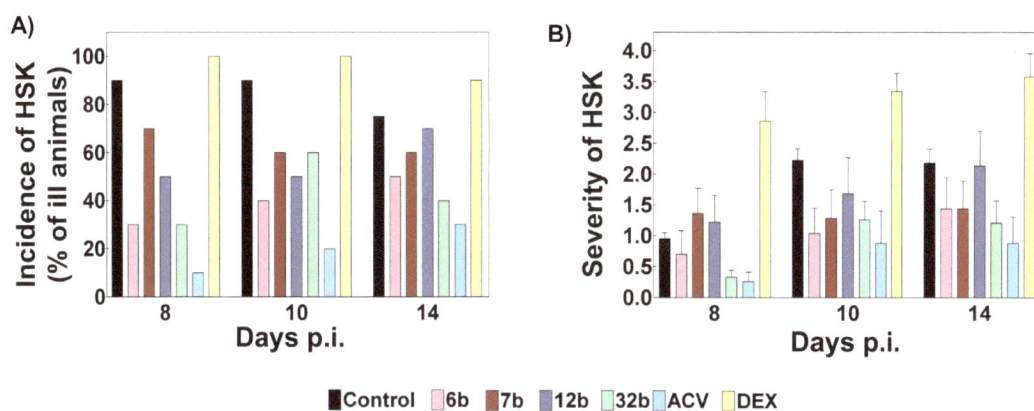

Figure 5: Antiviral effect of compounds **6b**, **7b**, **12b** and **32b** in the development of HSK. Mice corneas were inoculated with HSV-1 and treated with PBS (control), 40 µM of **6b**, **7b**, **12b** or **32b**, **ACV** (133 mM) and **DEX** (10 mM), three times a day, for three consecutive days starting on day 1 pos-infection (p.i.). Signs of keratitis (blepharitis, neovascularization, corneal opacity, edema, irritation and necrosis) were evaluated and registered during 14 days. **A)** percentage of sick animals (incidence of disease) and **B)** severity of the observed lesions.

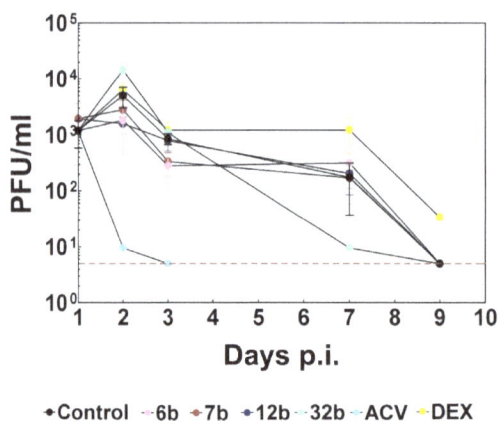

Figure 6: Effect of **6b**, **7b**, **12b** and **32b** on the multiplication of HSV-1 in the eye. Viral infectivity was titrated on the eye washes of the animals treated with PBS (control), **6b** (40 μM), **7b** (40 μM), **12b** (40 μM), **32b** (40 μM), **ACV** (133 mM) and **DEX** (10 mM).

Topic administration of 40 μM of BR **6b** and **32b** three times a day, on days 1, 2 and 3 p.i., significantly reduced the incidence of disease compared to control animals, which received PBS only (Fig. **5A**). Furthermore, **32b** was reduced the severity of the lesions in sick animals (Fig. **5B**). Both compounds were as effective as 133 mM of **ACV**, used as a positive control of specific antiviral activity. Since **6b** and **32b** did not exhibit antiviral effect *in vivo*, **6b** and **32b,** and **ACV** must heal HSK through different action mechanism (Fig. **7**) [16, 46]. A healing effect of compounds **7b** and **12b** was less evident. Although the percentage of sick animals in **7b** and **12b**-treated animals decreased, compared to untreated infected animals, differences were not statistically significant (Figs. **5A** and **B**) [16].

Even though viral replication occurred in the eyes of **6b** and **32b**-treated mice, unexpectedly, compounds **6b** and **32b** improved HSV-1-induced ocular disease, suggesting an eventual immunomodulating activity for those compounds. Compounds **6b** and **32b** significantly restrained the signs of keratitis when administered during the first three days p.i., while **DEX** exacerbated ocular damage (Figs. **5A** and **B**). In fact, after **DEX** treatment, score values largely surpassed those of the control, probably due to high viral loads that could not be completely eliminated until day 9 p.i. (Fig. **6**) [46]. In consequence, they would not behave as conventional steroidal anti-inflammatory drugs like **DEX,** which was included to evaluate its effect during the first stage of disease, when immunopathology had not been developed yet.

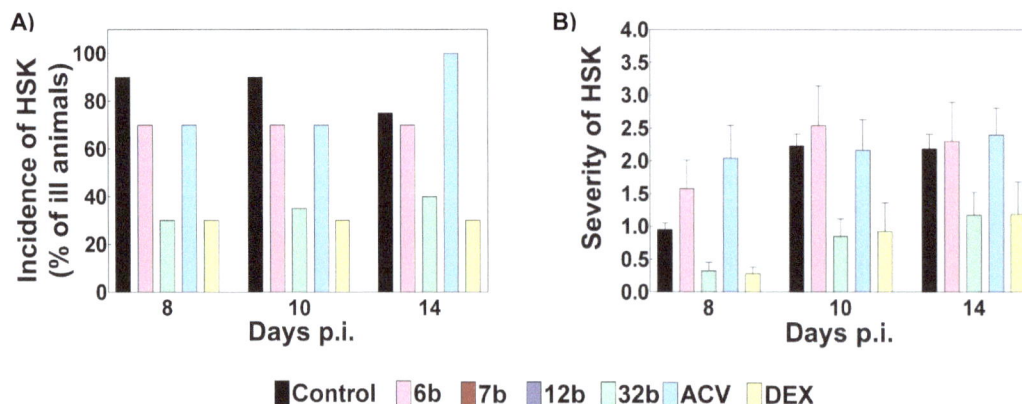

Figure 7: Anti-inflammatory effect of compounds **6b** and **32b** in the evolution of HSK. Mice corneas were inoculated with HSV-1 and treated with PBS (control), 40 μM of **6b**, **7b**, **12b** or **32b**, **ACV** (133 mM) and **DEX** (10 mM), three times a day, for three consecutive days starting on day 6 p.i. Signs of keratitis (blepharitis, neovascularization, corneal opacity, edema, irritation and necrosis) were evaluated and registered during 14 days. **A)** percentage of ill animals (incidence of disease) and **B)** severity of the observed lesions.

Assuming a potential anti-inflammatory activity for compounds **6b** and **32b**, those compounds were also assayed in the HSK model, using a different treatment schedule: compounds were provided topically for three consecutive days, starting on day 6 p.i., when typical lesions of disease begin to appear. The incidence of ocular disease remained low throughout the observation period, and a significant decrease in the severity of the lesions was observed in the group of animals treated with 40 μM of **32b**, with similar percentage and score values obtained after **DEX** administration (Figs. **7A** and **B**). However, no significant difference among untreated and **6b**-treated animals were registered (Figs. **7A** and **B**). As expected, **ACV** did not exert a healing effect when administered in this stage of disease, precisely because of the absence of infective virus in the corneas [16, 46].

In summary, to block recurrent episodes of HSK, an effective treatment is urgently needed. Ideally, this treatment would consist of: 1) an antiviral to block or significantly reduce neuronal activation of HSV-1, as well as viral anterograde transport from the neuron to the cornea; 2) an anti-inflammatory to reduce the corneal immune response but not to elicit an increase in ocular viral load; and 3) a drug to block or significantly reduce corneal neo-vascularization. However, this ideal combination therapy is not currently available.

The screening of the original BRs and the newly synthesized steroidal molecules for antiviral and anti-inflammatory activities yielded two stigmastanes which gather simultaneously the two activities. The identification of these drugs will help to overcome the current bottleneck existing in clinical treatments, *i.e.*, the possibility of the use of a single drug which fulfill the requirements of an anti-inflammatory regimen along with an antiherpetic program. If this dilemma can be overcome by the administration of these new BR analogs, the healing of immunopathologies like HSK will be on hand.

REFERENCES

[1]   Tyler KL, Nathanson N. Pathogenesis of viral infections. In : Knipe DM, Howley PM, Griffin DE, *et al.* Fields Virology. Philadelphia, PA, U.S.A.: Lippincott Williams and Wilkins, electronic version, 2001.

[2]   Bwaka MA, Bonnet MJ, Calain P, *et al.* Ebola hemorrhagic fever in Kikwit, Democratic Republic of the Congo: clinical observations in 103 patients. J Infect Dis 1999; 179 Suppl 1, S1-7.

[3]   Roizman B, Knipe DM, Whitley R J. Herpes simplex viruses. In: Knipe DM, Howley PM, Eds. Fields Virology. Philadelphia, PA, U.S.A.: Lippincott Williams & Wilkins 2007; pp. 2503-2602.

[4]   Knipe DM, Samuel CE, Palese P. Virus-Host cell interactions. In : Knipe DM, Howley PM, Griffin DE, Lamb RA, Martin, MA, Roizman B, Straus SE (Eds.), Fields Virology. Philadelphia, PA, U.S.A.: Lippincott Williams and Wilkins, electronic version, 2001.

[5]   Nevins JR. Cell transformation by viruses. In : Knipe DM, Howley PM, Griffin DE, Lamb RA, Martin MA, Roizman, B., Straus SE. (Eds.), Fields Virology. Philadelphia, PA, U.S.A.: Lippincott Williams and Wilkins, electronic version, 2001.

[6]   Azam F, Koulaouzidis A. Hepatitis B virus and hepatocarcinogenesis. Ann Hepatol 2008; 7: 125-9.

[7]   Fujinami RS, von Herrath MG, Christen U, Whitton JL. Molecular mimicry, bystander activation, or viral persistence: infections and autoimmune disease. Clin Microbiol Rev 2006; 19: 80-94.

[8]   Toma HS, Murina AT, Areaux RG Jr, *et al.* Ocular HSV-1 latency, reactivation and recurrent disease. Semin Ophthalmol 2008; 23: 249-73.

[9]   Biswas PS, Rouse BT. Early events in HSV keratitis-setting the stage for a blinding disease. Microbes Infect 2005; 7: 799-810.

[10]   Carr DJ, Harle P, Gebhardt BM. The immune response to ocular herpes simplex virus type 1 infection. Exp Biol Med (Maywood) 2001; 226: 353-66.

[11]   Deshpande SP, Zheng M, Lee S, Rouse BT. Mechanisms of pathogenesis in herpetic immunoinflammatory ocular lesions. Vet Microbiol 2002; 86 : 17-26.

[12]   Wickham S, Carr DJ. Molecular mimicry versus bystander activation: herpetic stromal keratitis. Autoimmunity 2004; 37: 393-7.

[13]   Knickelbein JE, Hendricks RL, Charukamnoetkanok P. Management of Herpes Simplex Virus Stromal Keratitis: An Evidence-based Review. Survey Ophthalmol 2009; 54: 226-34.

[14]   Hill JM, Clement C. Herpes Simplex Virus Type I DNA in human corneas: what are the virological and clinical implications? J Infect Dis 2009; 200: 1-4.

[15]   Ramírez JA, Teme Centurion OM, Gros EG, Galagovsky LR. Synthesis and bioactivity evaluation of brassinosteroid analogs. Steroids 2000; 65: 329-37.

[16]   Michelini FM, Ramírez JA, Berra A, Galagovsky LR, Alché LE. *In vitro* and *in vivo* antiherpetic activity of three new synthetic brassinosteroid analogues. Steroids 2004; 69: 713-20.

[17]   Talarico LB, Ramírez JA, Galagovsky LR, Wachsman MB. Structure-activity relationship studies in a set of new brassinosteroid derivatives assayed against herpes simplex virus type 1 and 2 in cell cultures. Medical Chem Res 2002; 11: 434-44.

[18]   Wachsman MB, Lopez EM, Ramírez JA, Galagovsky LR, Coto CE. Antiviral effect of brassinosteroids against herpes virus and arenaviruses. Antivir Chem Chemother 2000; 11: 71-7.

[19]   Wachsman MB, Ramírez JA, Galagovsky LR, Coto CE. Antiviral activity of brassinosteroids derivatives against measles virus in cell cultures. Antivir Chem Chemother 2002; 13: 61-6.

[20]   Wachsman MB, Castilla V, Talarico LB, Ramírez JA, Galagovsky LR, Coto CE. Antiherpetic mode of action of (22S,23S)-3β-bromo-5α,22,23-trihydroxystigmastan-6-one *in vitro*. Int J Antimicrob Agents 2004a; 23: 524-6.

[21]   Wachsman MB, Ramírez JA, Talarico LB, Galagovsky LR, Coto CE. Antiviral activity of natural and synthetic brassinosteroids. Curr. Med. Chem. - Anti-Infective Agents 2004b; 3: 163-79.

[22]   Labrie F, Luu-The V, Labrie C, *et al.* Endocrine and intracrine sources of androgens in women: inhibition of breast cancer and other roles of androgens and their precursor dehydroepiandrosterone. Endocr Rev 2003; 24: 152-82.

[23]   Nawata H, Yanase T, Goto K, Okabe T, Ashida K. Mechanism of action of anti-aging DHEA-S and the replacement of DHEA-S. Mech Ageing Dev 2002; 123: 1101-6.

[24]   Yoshida S, Honda A, Matsuzaki Y, *et al.* Anti-proliferative action of endogenous dehydroepiandrosterone metabolites on human cancer cell lines. Steroids 2003; 68: 73-83.

[25]   Ramírez JA, Bruttomesso AC, Michelini FM, Acebedo SL, Alché LE, Galagovsky LR. Syntheses of immunomodulating androstanes and stigmastanes: comparison of their TNF-alpha inhibitory activity. Bioorg Med Chem 2007; 15: 7538-44.

[26]   Gipson IK. The ocular surface: the challenge to enable and protect vision: the Friedenwald lecture. Invest Ophthalmol Vis Sci 2007: 48: 4390-98.

[27]   Akhtar J, Tiwari V, Oh MJ, *et al.* HVEM and nectin-1 are the major mediators of herpes simplex virus 1 (HSV-1) entry into human conjunctival epithelium. Invest Ophthalmol Vis Sci 2008; 49: 4026-35.

[28]   Liesegang TJ. Herpes simplex virus epidemiology and ocular importance. Cornea 2001; 20: 1-13.

[29]   Babu JS, Thomas J, Kanangat S, Morrison LA, Knipe DM, Rouse BT. Viral replication is required for induction of ocular immunopathology by herpes simplex virus. J Virol 1996; 70: 101-7.

[30]   Denizot F, Lang R. Rapid colorimetric assay for cell growth and survival. Modifications to the tetrazolium dye procedure giving improved sensitivity and reliability. J Immunol Methods 1986; 89: 271-7.

[31]   van Furth R, Cohn ZA, Hirsch JG, Humphrey JH, Spector WG, Langevoort HL. The mononuclear phagocyte system: a new classification of macrophages, monocytes, and their precursor cells. Bull World Health Organ 1972; 46: 845-52.

[32]   Hayashi K, Kurata T, Morishima T, Nassery T. Analysis of the inhibitory effect of peritoneal macrophages on the spread of herpes simplex virus. Infect Immun 1980; 28: 350-8.

[33]   Sit MF, Tenney DJ, Rothstein JL, Morahan PS. Effect of macrophage activation on resistance of mouse peritoneal macrophages to infection with herpes simplex virus types 1 and 2. J Gen Virol 1988; 69: 1999-2010.

[34]   Mogensen SC, Virelizier JL. The interferon-macrophage alliance. Interferon 1987; 8: 55-84.

[35]   Bauer D, Mrzyk S, van Rooijen N, Steuhl KP, Heiligenhaus A. Macrophage-depletion influences the course of murine HSV-1 keratitis. Curr Eye Res 2000; 20: 45-53.

[36]   Bauer D, Schmitz A, van Rooijen N, Steuhl KP, Heiligenhaus A. Conjunctival macrophage-mediated influence of the local and systemic immune response after corneal herpes simplex virus-1 infection. Immunology 2002; 107: 118-28.

[37]   Cheng H, Tumpey TM, Staats HF, van Rooijen N, Oakes JE, Lausch RN. Role of macrophages in restricting herpes simplex virus type 1 growth after ocular infection. Invest Ophthalmol Vis Sci 2000; 41: 1402-9.

[38]   Conboy IM, Manoli D, Mhaiskar V, Jones PP. Calcineurin and vacuolar-type H+-ATPase modulate macrophage effector functions. Proc Natl Acad Sci USA 1999; 96: 6324-9.

[39]   Kim SK, Shin MS, Jung BK, *et al.* Effect of dehydroepiandrosterone on lipopolysaccharide-induced interleukin-6 production in DH82 cultured canine macrophage cells. J Reprod Immunol 2006; 70: 71-81.

[40]   Kipper-Galperin M, Galilly R, Danenberg HD, Brenner T. Dehydroepiandrosterone selectively inhibits production of tumor necrosis factor alpha and interleukin-6 [correction of interlukin-6]   in astrocytes. Int J Dev Neurosci 1999; 17: 765-75.

[41]   Lavagno L, Gunella G, Bardelli C, *et al.* Anti-inflammatory drugs and tumor necrosis factor-alpha production from monocytes: role of transcription factor NF-kappa B and implication for rheumatoid arthritis therapy. Eur J Pharmacol 2004; 501: 199-208.

[42]   Liu T, Tang Q, Hendricks RL. Inflammatory infiltration of the trigeminal ganglion after herpes simplex virus type 1 corneal infection. J Virol 1996; 70: 264-71.

[43]   Hendricks RL. An immunologist's view of herpes simplex keratitis: Thygeson Lecture 1996, presented at the Ocular Microbiology and Immunology Group meeting, Cornea 1996; 16: 503-6.

[44]   Brandt CR, Coakley LM, Grau DR. A murine model of herpes simplex virus-induced ocular disease for antiviral drug testing. J Virol Methods 1992; 36: 209-22.

[45]   Pifarré MP, Berra A, Coto CE, Alche LE. Therapeutic action of meliacine, a plant-derived antiviral, on HSV-induced ocular disease in mice. Exp Eye Res 2002; 75: 327-34.

[46]   Michelini FM, Ramírez JA, Berra A, Galagovsky LR, Alché LE. Anti-herpetic and anti-inflammatory activities of two new synthetic 22,23-dihydroxylated stigmastane derivatives. J Steroid Biochem Mol Biol 2008; 111: 111-6.

CHAPTER 8

Anticancer Activities of Brassinosteroids

Lucie Hoffmannová[a], Jana Oklešťková[a], Jana Steigerová[b], Ladislav Kohout[a], Zdeněk Kolář[b] and Miroslav Strnad[a]

[a]*Laboratory of Growth Regulators, Faculty of Science, Palacký University & Institute of Experimental Botany ASCR, Šlechtitel 11, Olomouc, 783 71, Czech Republic and* [b]*Laboratory of Molecular Pathology, Department of Pathology, Faculty of Medicine, Palacký University, Hn votínská 3, 775 15 Olomouc, Czech Republic*

Abstract: Molecular and cellular effects of two groups of antiproliferative agents, natural brassinosteroids (BRs) and their synthetic derivatives, were examined in different human cancer cell lines and in primary endothelial cells *in vitro*. Natural and synthetic BRs caused growth inhibition, cell cycle arrest and initiation of apoptosis in many different cancer cell lines. The inhibition of proliferation and migration of human endothelial cells by BRs was demonstrated and evidences were obtained that BRs initiate cell death by apoptosis. And, analogues of BRs were found to be more effective than natural BRs. Observed inhibition of migration and tube formation demonstrated the antiangiogenic activity of BRs. These findings indicate a potential use of BRs in the prevention of metastasis development. Investigation of the mechanisms of action of BRs in human cancer and endothelial cells using cellular and molecular techniques indicated the possible involvement of steroid receptors in BR action. However, BRs were shown not to bind directly to steroid receptors which demonstrate that BRs act *via* steroid receptor-independent pathway(s). Concluding, BRs and their derivatives are capable to inhibit growth of several human cancer cell lines and to inhibit angiogenesis-like behaviour of primary endothelial cells *in vitro*, as well.

Keywords: Antiangiogenic activity, anticancer drugs, apoptosis, breast cancer, metastasis development, cell cycle arrest.

1. BRASSINOSTEROIDS

The phytohormones known as brassinosteroids (BRs) are low-molecular weight steroid compounds occurring in plants [1]. The first brassinosteroid $(22R,23R,24S)$-$2\alpha,3\alpha,22,23$-tetrahydroxy-24-methyl-B-homo-7-oxa-5α-cholestan-6-one (brassinolide) was isolated in 1979 from pollen of *Brassica napus* L. [2]. Up to date, more than 70 of these phytohormones have been discovered [3]. BRs have been detected and identified in many different plant species. They are ubiquitously distributed through the plant kingdom from lower to higher plants. BRs occur in most organs of the higher plant, including pollen, anthers, leaves, stems, roots, flowers, seeds and grain [4, 5].

BRs play an important role in hormone signalling in plants and in the physiological responses of plants to environmental stimuli in processes such as seed germination, growth, cell division and differentiation, root and stem elongation, bending, reproductive and vascular development, membrane polarization and proton pumping, source/sink relationships, disease resistance, modulation of stress and senescence [5-8].

The BRs are essential for many growth and development processes in plants. Contradictory reports on the effects of BRs and steroids on cell division in different plant species and cultured cell lines have been previously reported [1, 9, 10]. However, the effect of BRs on cell division in plants has been lately shown to be mainly promotive. BRs can mimic the effect of cytokinins on plant cell division: both BRs and cytokinins induce *cycD3* gene expression and promote cell division during the early phases in plant cell cultures, suggesting that BRs are rate-limiting factors in the induction of the cell cycle [11].

*****Address correspondence to Miroslav Strnad:** Laboratory of Growth Regulators, Faculty of Science, Palacký University & Institute of Experimental Botany ASCR, Šlechtitelů 11, Olomouc, 783 71, Czech Republic. E-mail: miroslav.strnad@upol.cz

2. ANTIPROLIFERATIVE ACTIVITIES OF BRASSINOSTEROIDS AND THEIR DERIVATIVES

Information about the effects of BRs and their synthetic analogues on animal and/or human cells is still very limited. However, recently, the first medical applications of BRs were published. These studies reported that some natural BRs, such as 28-homocastasterone (28-homoCS) or 28-homobrassinolide (Fig. 1) and their synthetic analogues, have *in vitro* antiviral activity against several pathogenic viruses [12-14]. There is also a report describing the effects of 24-epibrassinolide (24-epiBL) on cultured hybridoma mouse cells [15]. Typical effects of 24-epiBL were: (I) increase in mitochondrial membrane potential; (II) reduction in intracellular antibody level; (III) increase in number of cells in G_0/G_1 phase; (IV) and decrease the proportion of cells in S phase. Furthermore, the density of viable cells was significantly higher at 24-epiBL concentrations of 10^{-13} and 10^{-12} mol/l [15].

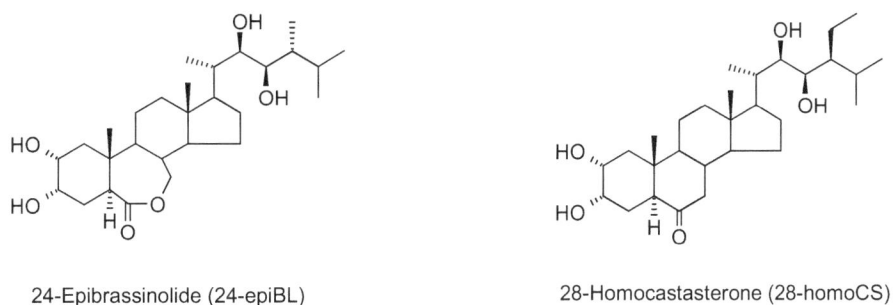

24-Epibrassinolide (24-epiBL)   28-Homocastasterone (28-homoCS)

Figure 1: Structures of 24-epibrassinolide and 28-homocastasterone.

An inhibitory effect of the natural BRs, 24-epiBL and 28-homoCS, on the growth and viability of different normal and cancer cell lines has been recently reported [16]. Using Calcein AM assay, 28-homoCS and 24-epiBL were shown to affect the viability of BJ fibroblasts and human cancer cell lines of various histopathological origins. Cell lines tested included: the T-lymphoblastic leukaemia CEM, breast carcinoma MCF-7, lung carcinoma A549, chronic myeloid leukaemia K562, multiple myeloma RPMI 8226, cervical carcinoma HeLa, malignant melanoma G361 and osteosarcoma HOS cell lines. Treatments with 28-homoCS and 24-epiBL resulted in potent, dose-dependent reductions in the viability of CEM and RPMI 8226 cells, albeit at different levels [16, 17]. Cytotoxicity of natural brassinosteroids against various human cancer lines is shown in Table **1**. The results provide several indications regarding structural features that are associated with cytotoxic activity. The most active compound against CEM cells was 28-homoCS, which induced approximately three times stronger responses than 28-homobrassinolide (28-homoBL), indicating that transformation of 6-oxo-7-oxalactone to 6-oxo functionality substantially increases the growth inhibitory activity of BRs; the presence of a 24R side chain strongly reduces BR cytotoxicity; the 24R side chain in 24-epicastasterone also reduces anticancer activity compared to castasterone. However, 28-homoCS and 28-homoBL, both of which have an ethyl group in their side chains at C24, are somewhat more effective than the corresponding analogues with C24 methyl groups. Since β-ecdysone, which contains 2β,3β,22α –functionality, showed no detectable activity, a 3α-hydroxy group, 2α,3α-vicinal diol or 3α,4α-vicinal diol may be important for cytotoxic activity [16]. Further testing of natural BRs (see Table **1**) such as typhasterol and teasterone, however, showed that these 3α-hydroxy analogues are inactive when tested on different cancer cell lines. Dolicholide type BRs also showed marginal anticancer activity. Many natural products were described as a source of new anticancer drugs. Up to date, more than 70 anticancer molecules are derived from natural products [18].

Estrogen- and androgen-sensitive and insensitive breast and prostate cancer cell lines were shown to respond differently to treatment with natural BRs. Most breast cancers consist of a mixture of estrogen-sensitive and estrogen-insensitive cells, and the key to the control of breast cancer seems to lie in the elimination of both cell types. Hormone-sensitive cell lines were more susceptible to treatment with BRs. This finding may point to possible modulation of steroid receptor-mediated responses by natural BRs. Furthermore, a cytotoxic effect of natural BRs was observed in cancer cells, but not in untransformed human fibroblasts, suggesting that BRs induce different responses in cancer and normal cells. Therefore, these plant hormones are promising candidates for development as potential anticancer drugs [16]. Brassinosteroids are also able to disturb cell cycling in breast and prostate cancer cell lines. Using flow cytometry, it was shown that treatment of breast and prostate cell lines with 28-homoCS and 24-epiBL blocked the cell cycle in the G_1 phase, with concomitant reductions in the percentages of cells in the S phase [16]. In the MCF-7 breast cancer cell model, the most widely used experimental system to study breast cancer, the typical growth inhibitory response of antiestrogens is manifested by similar reductions in the proportions of cells synthesizing DNA (S phase) after the antiestrogen treatment, and a corresponding increase in the proportions of cells in G_0/G_1 phase [19].

Table 1: Cytotoxicity of natural BRs against human cancer cell lines and human fibroblasts.

IC_{50}, μM	Cell line				
Compound	CEM	MCF-7	HeLa	A549	BJ
Cholesterol	>50	>50	>50	>50	>50
Castasterone	16.6 ± 5.3	>50	>50	>50	>50
Brassinolide	>50	>50	>50	>50	>50
28-Homobrassinolide	48.1 ± 1.3	>50	>50	>50	>50
28-Homocastasterone	13.2 ± 2.8	>50	>50	>50	>50
24-Epicastasterone	>50	>50	>50	>50	>50
24-Epibrassinolide	44.0 ± 2.2	>50	>50	>50	>50
Brassicasterol	>50	>50	>50	>50	>50
S,S-24-Epibrassinolide	>50	>50	>50	>50	>50
S,S-24-Epicastasterone	49.0 ± 1.8	>50	>50	>50	>50
S,S-Homobrassinolide	35.3 ± 2.2	>50	>50	>50	>50
S,S-Homocastasterone	24.4 ± 1.5	45.0±3.1	>50	>50	>50
Teasterone	>50	>50	>50	>50	>50
Homodolicholide	>50	>50	>50	>50	>50
Homodolichosterone	>50	>50	>50	>50	>50
Dolicholide	>50	>50	>50	>50	>50
Dolichosterone	>50	>50	>50	>50	>50
Typhasterol	43.0±9.3	37.9±13.0	>50	>50	>50

To determine the biological effects of synthetic cholestane derivatives of BRs (Fig. **2**) in cancer cells, various cell-based assays were used. In order to evaluate the cytotoxic properties of several cholestane derivatives and a related steroid, cholesterol, on the viability of normal and cancer cell lines of different histopathological origin, we used cells of a T-lymphoblastic leukemia cell line, CEM, a myeloma cell line, RPMI 8226, a breast carcinoma cell line, MCF-7, a cervical carcinoma cell line, HeLa, a human glioblastoma cell line, T98, and, as controls, normal human skin fibroblast cells, BJ. We observed a potent and dose- dependent decrease in the viability of CEM, RPMI 8226 and HeLa cells, albeit at different concentrations. The highest cytotoxicity was observed after application of cholestanon and compound 1966 (3beta-hydroxy-7a-homo-cholest-5-en-7a-one) [20]. Cholesterol, which is a non-cholestane derived plant and animal sterol, was, however, inactive or exhibited almost null cytotoxic activity. The 24R side chain was also shown to be a decisive group to increase the cytotoxicity of cholestane derivatives. Changing the 6-oxo-7-oxalactone to a 6-oxo functionality dramatically increased the growth inhibitory activity of the cholestane derivatives. In the BJ human fibroblasts, a cholestane derivative-mediated loss of viability was not observed. These results suggest that cancer cells and normal cells respond differentially to cholestane derivatives (manuscript in preparation). At present, only a few natural agents are known to posses the potential ability to selectively/preferentially eliminate cancer cells without affecting growth of normal cells. Results point to the potential use cholestane derivatives as pharmaceuticals for inhibition of hyperproliferation in tumors [21].

Because of the strong anticancer activities of the cholestane derivatives cholestanon and 1966, we further examined their antiproliferative properties. Flow cytometry analysis showed an increase in the proportion of cells in the subG$_1$ phase of the cell cycle (apoptotic cells) in MCF-7 and MDA-MB-468 cell lines after treatment with cholestanon or 1966. Cholestanon treatment increased the proportion of cells in the subG$_1$ and G$_0$/G$_1$ phase and decreased the proportion in the S phase. Treatment with 1966 caused a strong increase in the proportion of subG$_1$ phase cells (apoptotic cells) and a concomitant decrease in other cell phases. Treatment with cholestane derivatives resulted in decreased percentage of cells in the S phase in both, MCF-7 and MDA-MB-468 cell lines, being the effect stronger in MCF-7 estrogen-sensitive breast cancer cells [21].

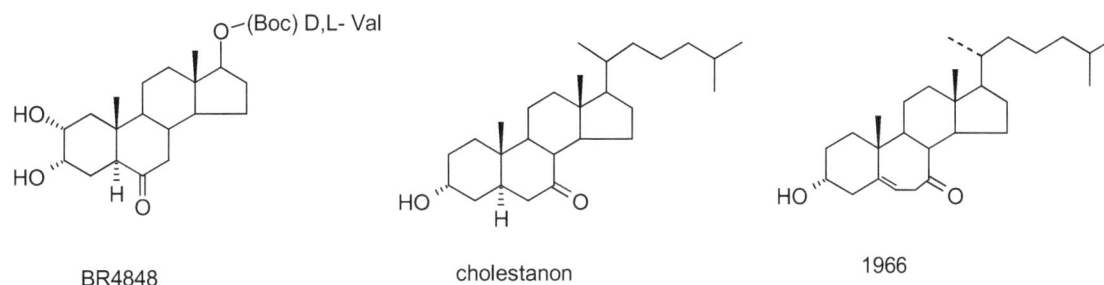

Figure 2: Structures of three synthetic derivatives of brassinosteroids.

Using a fluorogenic substrate Ac-DEVD-AMC and/or the caspase 3 and -7 inhibitor Ac-DEVD-DHO, the activity of caspases 3 and -7 in MDA-MB-468 cells exposed to cholestanon or 1966 was determined. Caspases 3 and -7 start the apoptotic cascade in cells. Cells treated with cholestanon presented a strong five-fold increase in the activity of effector caspases 3 and -7, when compared to untreated cells. However, derivative 1966 only weakly affected the activity of caspase 3 and -7.

To detect changes in apoptosis-related protein expression in breast cancer cell lines, cells treated with cholestane derivatives were used for Western blot immunodetection. Expression of a tumor suppressor protein, p53, was observed in controls of breast cancer cell lines, and, cholestanon and compound 1966 enhanced its expression. Enhanced expression of p53 correlated with a decreased expression of the antiapoptotic protein Mcl-1 in cells treated with cholestanon and compound 1966. These results confirm that the cholestane derivatives cholestanon and compound 1966 induce, in a dose-dependent way, apoptosis through caspase 3 and 7 activation [21].

3. ANTIANGIOGENIC PROPERTIES OF BRASSINOSTEROIDS AND THEIR ANALOGUES

Angiogenesis (Fig. **3**), the growth of new blood vessels in animals, is essential for organ growth [22-25] as well as for growth of solid tumors and for metastasis [25-27]. Endothelial cells are the main players in angiogenesis [27] and these cells could be a target for antiangiogenic therapy because they are non-transformed, and easily accessible to antiangiogenic agents. Endothelial cells are also unlikely to acquire drug resistance, because these cells are genetically stable, homogenous and have a low mutation rate [26]. The vascularization of tumors plays a crucial role in cell nutrition and oxygen distribution. Targeting tumor angiogenesis using novel drugs could potentially be achieved by the inhibition of proteolytic enzymes, which break down the extracellular matrix surrounding existing capillaries, and by inhibition of endothelial cell proliferation, migration and enhancement of tumor endothelial cell apoptosis. Potent angiogenic inhibitors, capable of blocking tumor growth, appear to have the potential for the development of novel generations of anticancer drugs [28-31].

Many natural products that inhibit angiogenesis have been reported, including compounds with steroid structure. These compounds include, among others, an active component of chilli peppers, capsaicin [32]; a low molecular weight natural product isolated from Dendrobium chrysotoxum Lindl, erianin [33]; a small molecule from extracts of the seed cone of Magnolia grandiflora L., honokiol [34]; a natural product from marine sponges, laulimalide [35]; the plant alkaloid, sanguinarine [36]; and a sesquiterpene purified from fruits of

Torilis japonica (Houtt.) DC., torilin [37]. Recently, several steroids, including 2-methoxyestradiol, progestin, medroxyprogesterone acetate, and glucocorticoids such as dexamethasone and cortisone (Fig. 4),

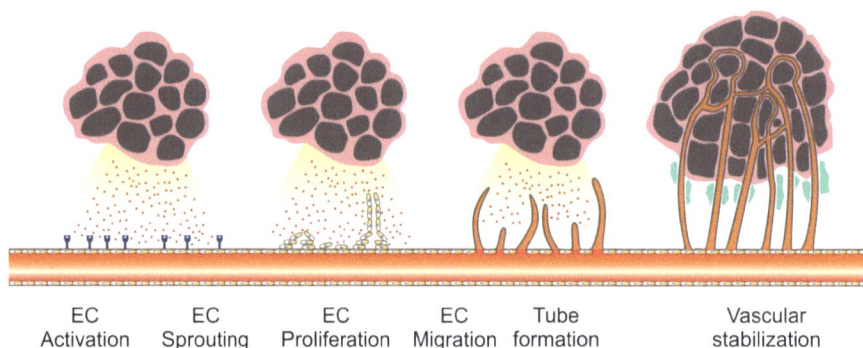

Figure 3: Angiogenesis in tumors. Angiogenesis in tumors is the creation of a network of blood vessels that penetrates into tumors. New blood vessels supply nutrients and oxygen and remove waste products. Starting molecules activate cancer cells which send signals to surrounding normal tissue to initiate tumor angiogenesis. This signaling activates certain genes in the normal endothelial cells (EC) to make proteins which stimulate growth of new blood vessels.

have been shown to present antiangiogenic activity [38, 39]. However, to date, there is no information on the effect of natural BRs on endothelial cells, including potential antiangiogenic activity. Therefore, we decided to investigate the effects of naturally occurring BRs and some of their synthetic analogues on cell proliferation and cycling in human microvascular endothelial (HMEC-1) and umbilical vein endothelial cells (HUVEC), more specifically on the migration and formation of tubes by these cells (submitted manuscript). All of the tested compounds, two natural brassinosteroids (24-epiBL, 28-homoCS) and two synthetic analogues of BRs (BR4848, cholestanon) (submitted manuscript), inhibited growth of HMEC-1 cells in a dose-dependent manner. 24-epiBL and 28-homoCS reduced the number of viable cells, while both synthetic analogues were similarly effective, but at a three-fold lower concentration. Both, BR4848 and cholestanon, decreased the number of cells adhering to a plastic surface more strongly than it did to a plastic surface coated with collagen. This effect may contribute to the antiproliferative activity observed for the tested compounds once different cells types require adhesion to a solid surface in order to proliferate (submitted manuscript).

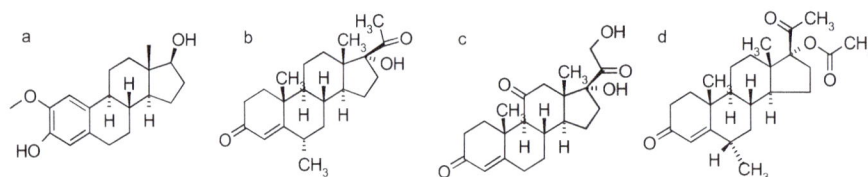

Figure 4: Structures of 2-methoxyestradiol (**a**), progestin (**b**), cortisone (**c**) and medroxyprogesterone acetate (**d**).

Flow cytometric analysis showed that treatment with 24-epiBL or 28-homoCS only slightly increased the proportion of cells in the subG$_1$ (apoptotic) fraction in HMEC-1 cells, when compared to the untreated controls. In contrast, treatment with synthetic derivatives enhanced the number of subG$_1$ cells, compared to the untreated controls. Moreover, BR4848 blocked the G$_2$/M phase of the cell cycle, similarly to what had been previously reported for 2-methoxyestradiol [40]. The tested steroids were more effective towards cell cycle arrest and induction of apoptosis than were the natural BRs, which only had cytostatic effects.

The two natural BRs tested, 24-epiBL and 28-homoCS, reduced migration of HUVEC cells, however, BR analogues presented a stronger inhibitory effect. Natural BRs caused only a slight inhibition of tube formation. Treatment with 24-epiBL or 28-homoCS slightly reduced the number of tubes, as well as the number of nodes, one of the parameters usually used to access antiangiogenic activity. Synthetic analogues also decreased the number of tubes compared to the control treatments (submitted manuscript). This antiangiogenic activity of BRs, along with their antiproliferative activity, suggests that these plant hormones might become important for the development of new anticancer drugs.

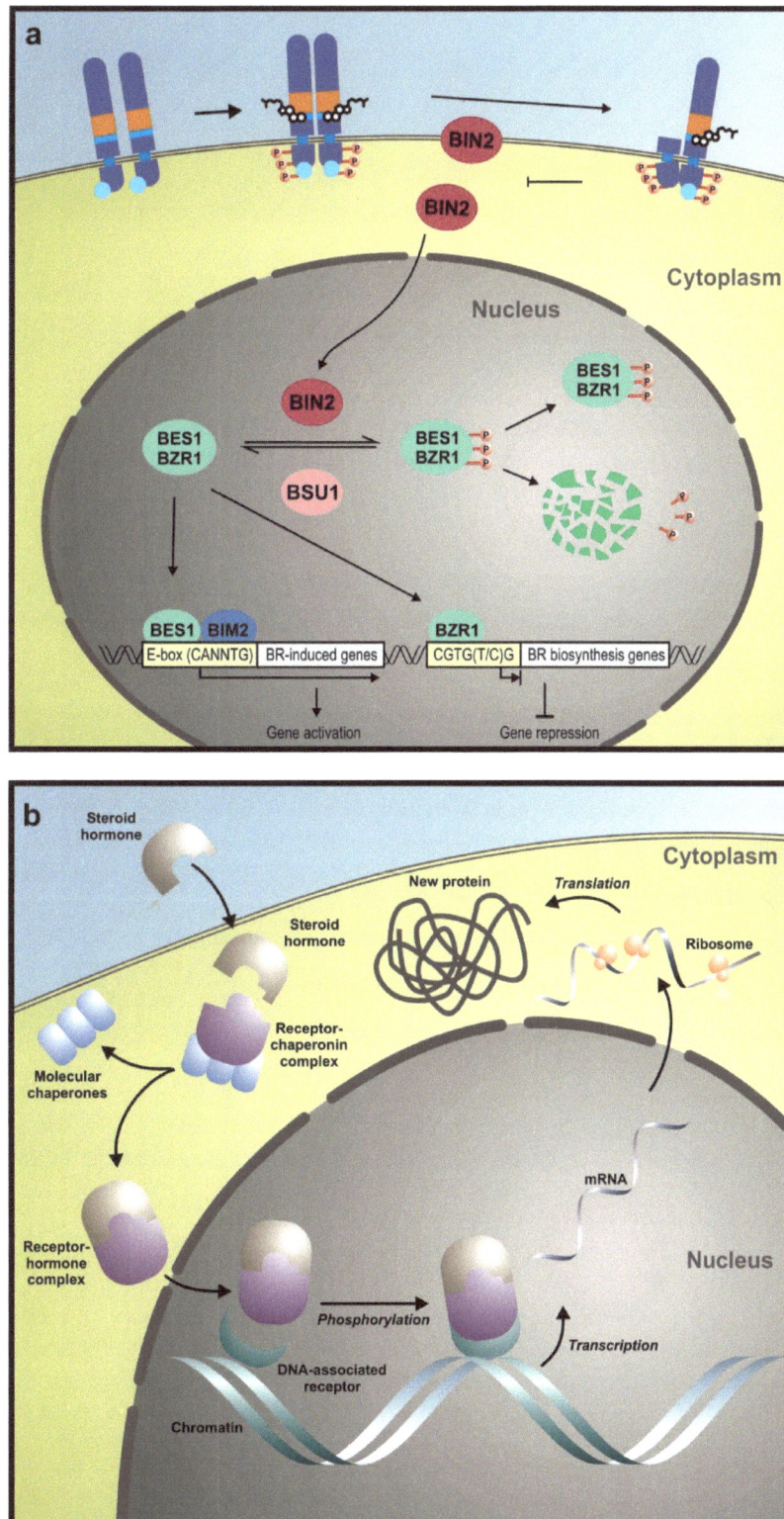

Figure 5: Actions of brassinosteroids in the plant cell **(a)** and possible action mechanism in animal cells **(b)**. (a) BRs bind to the extracellular domain of brassinosteroid-insensitive 1 (BRI1), a leucine-rich repeat receptor kinase (LRR-RK), localized in the plasma-membrane. This leads to phosphorylation of the BRI1 intracellular serine-threonine kinase domain, causing disassociation from the membrane-bound BRI1 kinase inhibitor 1 (BKI1) and oligomerization with a second receptor kinase, BRI1-associated receptor kinase 1 (BAK1). The active BRI1/BAK1 receptor kinase pair then propagates the signal downstream by inactivating a soluble kinase, brassinosteroid-insensitive 2 (BIN2), which is a

negative regulator of BR signaling. BES1 (bri1-EMS-suppressor 1) and BZR1 (brassinazole-resistant 1) are phosphorylated by BIN2 and are closely related transcriptional activators of BR-induced genes. BSU1 (bri1 suppressor 1) counteracts the effects of BIN2 [41]. (b) Steroid receptors (such as those for estrogen, androgen, progesterone, mineralcorticoid and glucocorticoid) are ligand-activated transcription factors that belong to the nuclear hormone receptor super-family. Lipophilic steroids that diffuse through the plasma membrane bind to the steroid receptors located in the cytosol or nucleus. Ligand binding induces a conformational change and dimerization with another receptor that allows the ligand/receptor complex to bind to the DNA and directly modify gene expression and protein synthesis [42].

4. MOLECULAR MECHANISM OF ACTION OF BRs

Signalling by BRs, and the resulting genomic responses in plants, are initiated by the binding of a BR molecule to a receptor kinase, brassinosteroid-insensitive 1 (BRI1), localized in the plasma-membrane [41, 42]. There are however differences between the action of steroids in animal cells and in plant cells (Fig. **5**). In the classic animal model, lipophilic steroids bind to steroid receptors located either in the cytosol or in the nucleus that diffuse through the plasma membrane. Ligand binding induces a conformational change and dimerization with another receptor that allows the ligand/receptor complex to bind to the DNA and directly modify gene expression over a time period of hours or even days [43].

Steroid hormone receptors exert their influence in embryonic development and adult homeostasis, as hormone-activated transcriptional regulators. Their modular structure, consisting of a DNA binding domain (DBD), nuclear localization signals, a ligand-binding domain (LBD) and several transcriptional activation functions, are conserved among other members of the nuclear receptor family. All unliganded steroid hormone receptors are associated with a large multiprotein complex of chaperones, including Hsp90 and the immunophilin Hsp56, which maintains the receptors in an inactive but ligand-friendly conformation, in contrast to other nuclear receptors such as glucocorticoid receptor [44]. Steroid receptors interact directly, *in vitro*, with components of the transcription initiation complex. The binding of ligands is influenced by co-activators that would act as bridging factors between steroid receptors and the transcription initiation complex. Steroid receptors not only stimulate gene activity, but they are also able to repress transcription by competition for the DNA-binding site, by competition for common mediators of the transcription initiation complex, or by sequestration of the transcription factors into inactive forms [45].

Because of the similarity between BRs and human steroids, we have also studied interactions of BRs with human steroid receptors using reporter assays and a competition binding assay. Reporter assays showed that 24-epiBL was a weak antagonist of estrogen-receptor-α (ER-α); the synthetic BRs, cholestanon and BR4848, showed agonistic effects on ER-α, estrogen-receptor-β (ER-β) and androgen receptor (AR). Despite the results of the reporter assays, we were unable to demonstrate direct binding of the tested steroids to ER-α and ER-β in competition binding assays, with the exception of cholestanon, which showed weak binding to both, ER-α and ER-β. 24-epiBL, 28-homoCS and BR4848 did not bind to estrogen receptors. Summarizing, all of the tested compounds, in one or another way, showed effects in our angiogenesis assays *in vitro*, at micromolar concentrations. Furthermore, the ability of the tested compounds to modulate steroid responses in reporter cell lines was significantly strong. On the other hand, direct binding of the tested compounds, at least to steroid receptors ER-α and ER-β, is weak. Therefore, we can only speculate about the action mechanism of natural BRs and synthetic derivatives on angiogenesis. It seems possible that there are multiple effects, both steroid receptor-dependent and independent, similarly to what is considered for 2-methoxyestradiol, which binds to estrogen receptors. However, the action mechanism of the antiangiogenic response is probably unrelated to the receptor pathway (submitted manuscript).

On the other hand, changes in pattern of ER-α and ER-β localization were observed in MCF-7 cells, after BR treatment, using immunofluorescence detection. Strong, uniform ER-α immuno-nuclear labeling was detected in untreated MCF-7 cells, while cytoplasmic speckles of ER-α immunofluorescence were observed in MCF-7 cells treated with 28-homoCS or 24-epiBL. In contrast to ER-α, ER-β was predominantly found in the cytoplasm of untreated MCF-7 cells. However, ER-β was notably re-located to the nuclei after 28-homoCS treatment, whereas it was predominantly present at the periphery of the nuclei, in 24-epiBL-treated cells. These changes were also accompanied by down-regulation of the ERs following BR treatment (submitted manuscript). The studied BRs seem to exert potent growth inhibitory effects *via* interactions with the cell cycle machinery, and they could be valuable candidates for agents to manage breast cancer.

Our work has shown that the BRs 28-homoCS and 24-epiBL have dose-dependent effects on the viability of estrogen-sensitive MCF-7 cells and estrogen-insensitive MDA-MB-468 cells. In breast cancer cells, BRs inhibited cell growth and blocked the G_1 phase of the cell cycle, with concomitant reductions in the percentages of cells in the S phase. Cell cycle arrest was accompanied by reductions in cyclin-dependent kinases (CDKs) 2/4/6, cyclin D_1 and E expression and pRb phosphorylation, together with up-regulation of the cyclin-dependent kinase inhibitors p21$^{Waf1/Cip1}$ and p27^{Kip1}, which inhibit cyclin/CDK complexes. In addition, 28-homoCS and 24-epiBL treatments induced expression of the anti-apoptotic Bcl-2 and Bcl-X$_L$ proteins in MCF-7 cells. The levels of p53 and MDM-2 proteins were slightly affected by BR treatments. In MDA-MB-468 cells it was found that caspase-3 was cleaved into fragments (part of the apoptotic cascade) 24 h after the BR treatment. Furthermore, BR application to MDA-MB-468 cells resulted in G_1 phase arrest and increases in the subG$_1$ fraction, which represents apoptotic bodies. It was confirmed that the BR-mediated apoptosis occurred in both cell lines, although changes in the expression of apoptosis-related proteins were modulated differently by the BRs in each cell line [46]. When seem together, these data indicates that a wide range of carcinomas are likely to be targets for the antiangiogenic effects of BRs and their synthetic derivatives. The action mechanisms of BRs in animal cells are still largely unknown, but it seems possible that BRs may interact with one or more of the numerous steroid-binding proteins. It also seems possible that BRs induce multiple effects, both, steroid receptor-dependent and independent. These properties may lead to the development of novel natural product-derived anticancer drugs.

ACKNOWLEDGEMENTS

The authors wish to thank Prof. D.A. Morris (UK) for his suggestions and critical reading of the manuscript. This work was supported by the Ministry of Education, Youth and Sports of the Czech Republic MSM6198959216 and IAA400550801.

REFERENCES

[1]   Khripach VA, Zhabinskii VN, de Groot AE. Brassinosteroids. A new class of plant hormones. San Diego: Academic Press 1999; 456p.

[2]   Grove MD, Spencer GF, Rohwedder WK, *et al.* Brassinolide, a plant growth-promoting steroid isolated from *Brassica napus* pollen. Nature 1979; 281: 216-7.

[3]   Sakurai A, Fujioka S. Studies on biosynthesis of brassinosteroids. Biosci Biotechnol Biochem 1997; 61: 757-62.

[4]   Bajguz A, Tretyn A. The chemical characteristic and distribution of brassinosteroids in plants. Phytochemistry 2003; 62: 1027-46.

[5]   Bajguz A, Hayat S. Effects of brassinosteroids on the plant responses to environmental stresses. Plant Physiol Biochem 2009; 47: 1-8.

[6]   Clouse SD, Sasse JM. Brassinosteroids: Essential Regulators of Plant Growth and Development. Annu Rev Plant Physiol Plant Mol Biol 1998; 49427-51.

[7]   Haubrick LL, Assmann SM. Brassinosteroids and plant function: some clues, more puzzles. Plant Cell Environ 2006; 29: 446-57.

[8]   Nemhauser JL, Chory J. BRing it on: new insights into the mechanism of brassinosteroid action. J Exp Bot 2004; 55: 265-70.

[9]   Jiang F, Li P, Fornace AJ, Jr., Nicosia SV, Bai W. G2/M arrest by 1,25-dihydroxyvitamin D3 in ovarian cancer cells mediated through the induction of GADD45 *via* an exonic enhancer. J Biol Chem 2003; 278: 48030-40.

[10]   Jun DY, Park HS, Kim JS, *et al.* 17Alpha-estradiol arrests cell cycle progression at G2/M and induces apoptotic cell death in human acute leukemia Jurkat T cells. Toxicol Appl Pharmacol 2008; 231: 401-12.

[11]   Hu Y, Bao F, Li J. Promotive effect of brassinosteroids on cell division involves a distinct CycD3-induction pathway in *Arabidopsis*. Plant J 2000; 24: 693-701.

[12]   Romanutti C, Castilla V, Coto CE, Wachsman MB. Antiviral effect of a synthetic brassinosteroid on the replication of vesicular stomatitis virus in Vero cells. Int J Antimicrob Agents 2007; 29: 311-6.

[13]   Wachsman MB, Lopez EM, Ramirez JA, Galagovsky LR, Coto CE. Antiviral effect of brassinosteroids against herpes virus and arenaviruses. Antivir Chem Chemother 2000; 11: 71-7.

[14]   Wachsman MB, Ramirez JA, Galagovsky LR, Coto CE. Antiviral activity of brassinosteroids derivatives against measles virus in cell cultures. Antivir Chem Chemother 2002; 13: 61-6.

[15]   Franěk F, Eckschlager T, Kohout L. 24-Epibrassinolide at subnanomolar concentrations modulates growth and production characteristic of a mouse hybridoma. Collect Czech Chem Commun 2003; 68: 2190-2200.

[16]   Malikova J, Swaczynova J, Kolar Z, Strnad M. Anticancer and antiproliferative activity of natural brassinosteroids. Phytochem 2008; 69: 418-26.

[17]   Oklešťková J, Hoffmannová L, Steigerová J, Kohout L, Kolář Z, Strnad M. Natural brassinosteroids for use for treating hyperproliferation, treating proliferative diseases and reducing adverse effects of steroid dysfunction in mammals, pharmaceutical composition and its use.US Patent 20100204460, 2010.

[18]   Newman DJ, Cragg GM. Natural products as sources of new drugs over the last 25 years. J Nat Prod 2007; 70: 461-77.

[19]   Parl FF. Estrogens, Estrogen Receptor and Breast Cancer. Amsterdam: IOS Press 2000; 263p.

[20]   Kohout L, Fajkoš J, Šorm F. On Steroids CXXI. B-Homosteroids II. Conformation of the Ring B in B-Homosteroid Alcohols and Epoxides. Coll Czech Chem Commun 1969; 34: 601.

[21]   Hoffmannová L, Steigerová J, Oklešťková J, *et al.* Cholestane derivatives as antiproliferative and antiangiogenic pharmaceuticals and pharmaceutical preparations containing these compounds. CZ patent application CZ PV 2009-725. 2009.

[22]   Carmeliet P. Angiogenesis in life, disease and medicine. Nature 2005; 438: 932-6.

[23]   Folkman J. Tumor angiogenesis. Adv Cancer Res 1974; 19: 331-58.

[24]   Folkman J, Shing Y. Angiogenesis. J Biol Chem 1992; 267: 10931-4.

[25]   Bhat TA, Singh RP. Tumor angiogenesis--a potential target in cancer chemoprevention. Food Chem Toxicol 2008; 46: 1334-45.

[26]   Risau W. Mechanisms of angiogenesis. Nature 1997; 386: 671-4.

[27]   Bagley RG, Walter-Yohrling J, Cao X, *et al.* Endothelial precursor cells as a model of tumor endothelium: characterization and comparison with mature endothelial cells. Cancer Res 2003; 63: 5866-73.

[28]   Gourley M, Williamson JS. Angiogenesis: new targets for the development of anticancer chemotherapies. Curr Pharm Des 2000; 6: 417-39.

[29]   Ranieri G, Gasparini G. Angiogenesis and angiogenesis inhibitors: a new potential anticancer therapeutic strategy. Curr Drug Targets Immune Endocr Metabol Disord 2001; 1: 241-53.

[30]   Singh RP, Agarwal R. Tumor angiogenesis: a potential target in cancer control by phytochemicals. Curr Cancer Drug Targets 2003; 3: 205-17.

[31]   Kesisis G, Broxterman H, Giaccone G. Angiogenesis inhibitors. Drug selectivity and target specificity. Curr Pharm Des 2007; 13: 2795-809.

[32]   Min JK, Han KY, Kim EC, *et al.* Capsaicin inhibits *in vitro* and *in vivo* angiogenesis. Cancer Res 2004; 64: 644-51.

[33]   Gong YQ, Fan Y, Wu DZ, Yang H, Hu ZB, Wang ZT. *In vivo* and *in vitro* evaluation of erianin, a novel anti-angiogenic agent. Eur J Cancer 2004; 40: 1554-65.

[34]   Bai X, Cerimele F, Ushio-Fukai M, *et al.* Honokiol, a small molecular weight natural product, inhibits angiogenesis *in vitro* and tumor growth *in vivo*. J Biol Chem 2003; 278: 35501-7.

[35]   Lu H, Murtagh J, Schwartz EL. The microtubule binding drug laulimalide inhibits vascular endothelial growth factor-induced human endothelial cell migration and is synergistic when combined with docetaxel (taxotere). Mol Pharmacol 2006; 69: 1207-15.

[36]   Eun JP, Koh GY. Suppression of angiogenesis by the plant alkaloid, sanguinarine. Biochem Biophys Res Commun 2004; 317: 618-24.

[37]   Kim MS, Lee YM, Moon EJ, Kim SE, Lee JJ, Kim KW. Anti-angiogenic activity of torilin, a sesquiterpene compound isolated from *Torilis japonica*. Int J Cancer 2000; 87: 269-75.

[38]   Mabjeesh NJ, Escuin D, LaVallee TM, *et al.* 2ME2 inhibits tumor growth and angiogenesis by disrupting microtubules and dysregulating HIF. Cancer Cell 2003; 3: 363-75.

[39]   Pietras RJ, Weinberg OK. Antiangiogenic steroids in human cancer therapy. Evid Based Complement Alternat Med 2005; 2: 49-57.

[40]   Lee YM, Ting CM, Cheng YK, *et al.* Mechanisms of 2-methoxyestradiol-induced apoptosis and G2/M cell-cycle arrest of nasopharyngeal carcinoma cells. Cancer Lett 2008; 268: 295-307.

[41]   Fellner M. Recent progress in brassinosteroid research: Hormone perception and signal transduction. Amsterdam: Kluwer Academic Publishers 2003; 246p.

[42]   Belkhadir Y, Chory J. Brassinosteroid signaling: a paradigm for steroid hormone signaling from the cell surface. Science 2006; 314: 1410-1.

[43] Losel R, Wehling M. Nongenomic actions of steroid hormones. Nat Rev Mol Cell Biol 2003; 4: 46-56.

[44] Pratt WB. The role of heat shock proteins in regulating the function, folding, and trafficking of the glucocorticoid receptor. J Biol Chem 1993; 268: 21455-8.

[45] Beato M, Herrlich P, Schutz G. Steroid hormone receptors: many actors in search of a plot. Cell 1995; 83: 851-7.

[46] Steigerová J, Oklešťková J, Levková M, Rárová L, Kolář Z, Strnad M. Brassinosteroids cause cell cycle arrest and apoptosis of human breast cancer cells. Chem Biol Interact. 2010, 88(3): 487-96.

Index

A

Aluminium stress 46,51
Antiangiogenic activity 84, 87-91
Anticancer 5-6, 10, 85-88, 91
Antigenotoxicity 10
Anti-herbivore 36-38
Antiherpetic activity 68, 81
Apoptosis 4-6, 72, 84-91
Arena vírus 4-6

B

Brassica napus 3, 5, 16-17, 44, 84
Brassinazole 4, 90
Brassinolide 3-4, 8, 16, 27, 31, 44-45, 54, 57-60, 84, 86
BRI1 receptor 4, 37, 45
Brz 220 26, 30-31

C

Cadmium stress 46-53
Cancer 3, 5-7, 84-91
Caspase 87, 91
Castasterone 3, 30-31, 54, 85-86
Catalase 19, 53
Cell cycle 6, 72, 84-91
Copper 52-54
Cyclins 6

D

DNA fragmentation 4
Drought resistance 17-18

E

24-Epibrassinolide 5, 9, 18, 45-49, 51-52, 54, 85-86
24-Epicastasterone 85-86
Ecdysteroid 4
Ethylene 9, 17-20, 28-29, 36, 44
Eucalyptus 8, 19, 26, 31-32

F

5α-Fluorohomocastasterone 26-27

H

28-Homobrassinolide 5-6, 18, 45-54, 85-86
28-Homocastasterone 6, 26-27, 31, 47, 58-66, 85-86
Herbicides 7
Herpes simplex virus 4-6, 57, 72
Human acute lymphoblastic leukemia 6
Human breast adenocarcinoma 6
Human myeloma 6

www.ingramcontent.com/pod-product-compliance
Lightning Source LLC
Chambersburg PA
CBHW041720210326
41598CB00007B/726

9 781608 056545